THE LAB

THE LAB
Creativity and Culture

David Edwards

Harvard University Press
Cambridge, Massachusetts
London, England
2010

Library of Congress Cataloging-in-Publication Data
Edwards, David A., 1961–
    The lab : creativity and culture / David Edwards.
        p.   cm.
    Includes index.
    ISBN 978-0-674-05719-7 (alk. paper)
    1. Creative ability in science.   2. Laboratories.   3. Research, Industrial—
Laboratories.   4. Art and science.   5. Science in popular culture.   I. Title.
    Q172.5.C74E38 2010
    153.3′5—dc22   2010014196

Contents

1

DREAMS

Dreaming compels the passion of children and visionaries alike. It is what allows us, at a young age, to imagine the long years of schooling and apprenticeship that lie ahead, without abandoning the optimism and curiosity of childhood. It is also what allows us to pursue ideas that no one else dares imagine, in the face of convention, institutional rigidity, and doubt.

To dream is to believe in the possibility of experimentation. If we dream about swimming a great distance, we are not perfectly certain we can swim so far, not sure what obstacles we will face or how we will surmount them. But we do believe in the possibility of the experiment. Nature—the lake, the pool, the rushing river—becomes our lab, our place of experimentation.

Labs obviously come in various kinds—film labs, design labs, molecular biology labs. We do experiments in labs that aim at outcomes, which we often call discoveries. We can't say with certainty what these discoveries will be, nor can we determine when we'll achieve them, but we can imagine that film labs will mostly make discoveries of relevance to films, design labs to design, and molecular biology labs to molecular biology. This possibility of discovery encourages investment: by creators, wishing to sustain their dreams; by labs, aiming to sustain creators; and by supporters of labs, hoping to sustain innovation in a world that is constantly in flux.

This book is about a special kind of lab where creators and

society meet through cultural exhibition, a new platform for innovation.

≡≡

We dream, and realize dreams, through a creative process that mixes two ways of thinking—aesthetic and analytical—which we often encourage and exploit in very different settings. Through aesthetic thinking, we embrace uncertainty and complexity, we induce, follow intuition, and draw inspiration from images and sounds. This process especially thrives in artistic environments, like theater companies or design studios. Through analytical thinking, we simplify a complex world, reduce its challenges to resolvable problems, and pursue the logic of equations. This process thrives in scientific environments, like a pharmaceutical company or a bank. The aesthetic process is the substance of hypothesis generation, while the analytical process is the substance of hypothesis testing. Inevitably we fuse both when we create anything new. We dream and we analyze, we induce and we deduce, we embrace complexity and we simplify our world to problems we can solve. This fused process of artscience, while at the core of creativity, often gets left out of those specialized environments where we hope creativity will reign.

In our personal lives it is different. We have ideas and we dream—to become a dancer, a teacher, a skilled card player. Meanwhile, we busy ourselves with the development of these ideas—with idea translation. We hypothesize that by doing certain things with our lives—getting up in the morning, going to school, pursuing a degree, practicing a skill—we will move nearer to the realization of our dreams. When we do these things, in a sense we are testing our hypotheses. And, if we are lucky, we have other people in our lives who help us assess our progress along the way. Through dinner parties, phone conversations, lunch breaks, or perhaps long weekends with a partner in the country, we come to value the experimental process of idea translation even while we are still far short of realizing our dreams. All this helps us take stock of where we are, now—usually far from where we ultimately hope to be, but almost always in a pretty worthy place nevertheless.

We seek out such environments to curate the creative process of our lives. These private labs, which we build and re-build with friends and family, do not preclude wild hypotheses any more than they preclude careful analysis of what is likely to work. They do not filter out surprising inductive associa-tions any more than they prevent us from sharing penetrating deductive insights. Friends and family help us best when they listen to us, talk with us, and accompany us emotionally wher-ever the creative process of our lives happens to take us. This accompaniment gives us the energy and perspective to trans-late our ideas and realize our dreams.

The creative process of idea translation is hard to share, understand, or empathize with when we are not intimate par-ticipants. It is always easier to point to what a process pro-duces—a finished painting or a new drug or a sleek cell phone design. But inferring process from outcome is hard, generally impossible.

One afternoon let's say you visit the National Galleries in the Grand Palais in Paris and happen to see the nearly one hundred art objects, mostly paintings, which make up the blockbuster *Picasso and the Masters* retrospective. You spend ninety minutes inside the gallery, so you estimate that you spent, on average, less than a minute standing in front of each art object. This would be typical of a visit to an art exhibition: published statistics show that we spend on average three seconds before a work of art in a museum. So, in this sense at least, you have an ordinary visit inside an extraordinary exhibition. You remember certain paintings. For example, the connection between Picasso's work and Goya's earlier work really strikes you. You will speak about it when you're finished. And indeed, when someone casually asks you, in the days that follow, if you enjoyed *Picasso and the Masters,* and ask (when you say you did) what most impressed you about the retrospective, you will call on these details. They are what your visit was all about.

Now, in fact, your visit to the museum was much richer than this. It happens that the exhibition was oversold and nearly impossible to enter. A friend associated with the museum helped you circumvent the interminable lines and slip inside. While strolling through the galleries with this friend, you met three acquaintances, and, shockingly, one old friend. You stepped to the side and spoke with this old friend for several minutes, no more, but she stayed on your mind for the remainder of the visit. You couldn't forget how the last time you saw her she'd been a promising young art student at the Louvre with unkempt hair and worn jeans, angling for a brilliant curator's career. And now here she was, married, her hair trimmed neatly above the shoulder, living in Versailles with her young family. Nothing about her revealed the other person she had once been. And this encounter with your old friend, these memories, make you especially sensitive to Picasso's sometimes playful, sometimes critical, generally irreverent perspective on the time invariance of cultural norms.

So your visit goes something like this. It is a personal drama that passionately engages you, but you will not speak about it to just anyone, will not share the subtle details, perhaps judging it would take too long or that nobody outside a certain intimate circle would understand anyway without knowing many other things that would be too tiring for you to convey. So you stick to certain facts, the "outcome" of your visit (the connection between Picasso and Goya), and leave the details for those few who are familiar with your past and would appreciate the backstory of your day at the museum.

To remain curious, attentive, and moved by the rich process of idea translation, we seek out those from whom we can learn and with whom we can discuss our experiences and dreams. The translation of an idea from our early perception of a dream to a realized outcome typically follows an unpredictable path that requires nearly constant attention and dialog. Hence our need for supportive family, friends, and colleagues who will help us understand, accept, and even celebrate the experimental substance of our daily creative process.

This personal perspective on creativity is completely at odds with our collective experience. We value our lives for the creative process that infuses them, while we value what we call labs—that is, environments intended to curate creative process in order to benefit creators, investors, and society—for the product that comes out. This is not a good recipe for innovation.

In this book, I argue for a lab that improves the dialog

between creators and the public around the creative process while erasing conventional boundaries between art and science. This kind of artscience lab resembles the early twentieth-century German art school known as the Bauhaus, where ideas in the arts and design advanced through phases of experiential learning, cultural exhibition, and production, with students working alongside master artists and designers. It is illustrated today by several artscience centers that have appeared in recent years in London, Dublin, Paris, and several other cities.

The innovation model of these artscience labs, and particularly that of the art and design lab in Paris called Le Laboratoire, shares common features with other successful models for innovation, including those of the MIT Media Lab, Google, Ideo, and Ars Electronica Futurelab. The uniqueness of the artscience model relates principally to the centrality of cultural exhibition as a forum for idea expression, and to a multi-lab structure that funnels innovative ideas toward educational, cultural, commercial, and humanitarian benefits.

≡≡

Le Laboratoire opened in the fall of 2007. Earlier that summer the historian of science Ken Arnold opened the Wellcome Collection in London, and in January 2008 the art historian Michael John Gorman opened the Science Gallery in Dublin. These three cultural centers had a common mission in that they each brought artistic and scientific ideas, themes, and experimenters before the public in what seemed to be a new category of cultural institution. They were neither science centers nor art galleries but something in between.

These organizations worked to varying degrees with institutional partners. Ken led an initiative of the Wellcome Trust, and Michael John did the same for Trinity College, Dublin. Le Laboratoire emerged as a private organization with a strong educational link to Harvard University. These partners, to varying degrees, influenced the visions of their respective art-science organizations and made the labs' founding teams sensitive to the particular needs of large specialized institutions.

With articles appearing in the British journal *Nature* over the course of 2007 and 2008, the founders of the Dublin, London, and Paris labs began a conversation that grew amidst an enlarging network of vibrant artscience initiatives. These included The International Society for Arts, Science & Technology and its flagship journal *Leonardo*, founded in 1968 in Paris by the kinetic artist and scientist Frank Malina; Art & Sciences Collaborations, Inc. (ASCI), founded in 1988 by artist Cynthia Pannucci in New York to encourage support for technology-based art; SymbioticA, founded in 2000 at the University of Western Australia by cell biologist Miranda Grounds, neuroscientist Stuart Bunt, and artist Oron Catts as a permanent art lab working on biological science projects; the UCLA Art/Sci Center, founded in recent years to foster collaborations at UCLA between media arts and bio- and nanosciences; and the École des Arts Politiques, founded in 2009 by French philosopher Bruno Latour at Sciences Po in Paris.

These and other new organizations were hardly the result of a coordinated movement, or some kind of clearly recognized paradigm. Each organization had its own story, its own intrinsic ability to scale, as if some new idea had suddenly gone viral. Ken Arnold envisioned an art and science exhibition program featuring a vast collection of medical artifacts that raised contemporary questions about health and society, a core interest of the Wellcome Trust. Michael John Gorman seized on the desire of leaders within Trinity College to bring the Irish public into personal relationships with nanotechnology and other frontiers of scientific research by organizing art and design exhibitions in the sciences. And Bruno Latour aimed to broaden interdisciplinary thinking within the political sciences through artistic creation.

Le Laboratoire grew within the broad outlines of the science lab. But unlike its traditional counterparts, this new lab would host experiments to test and develop ideas in the visual and performing arts and in design. Exhibiting works-in-progress through a kind of peer review process, it would do whatever was required to translate ideas further into cultural, commercial, educational, and humanitarian benefits. But it quickly became clear that Le Laboratoire, with its particular model, could not exist in isolation for long. Other labs had to be put in place upstream and downstream of the idea translation process to help innovations emerge. In the same way that the contemporary science lab innovates in partnership with the university on one side and industry on the other, Le Laboratoire needed independent artscience labs focused on educational, humanitarian, and commercial idea translation.

Over the course of 2005–2010 we opened new labs on three continents: in Boston and Cambridge, in Paris, and in Pretoria and Cape Town. We soon began to think of this network of labs as an idea funnel, like the finely polished funnel that guides hypothetical ideas through peer-reviewed publication in science labs from their origins in academic educational settings to technological innovations in commercial settings. In the artscience lab model, art and design ideas would move from education on the one side to social and cultural change on the other, with public dialog taking place in between, through cultural exhibition instead of academic publication.

To understand how this idea funnel actually works, you might imagine a group of mostly African students in the educational artscience lab at Harvard University coming up with a new design approach to lighting villages in sub-Saharan Africa using microbial fuel cells. Following a public presentation at Harvard, the idea matures through an innovation workshop at the cultural artscience lab, Le Laboratoire, the next summer in Paris, where the students exhibit their idea before a public that includes Michael John Gorman of the Science Gallery. Michael John invites the students to Dublin, where they exhibit their idea again, helping it to advance further through public feedback. Meanwhile, the students receive a World Bank Award to develop a new nonprofit organization called Lebone devoted to realizing their dream of bringing light to African villages.

Ideas develop in this way, often originating at the top of the funnel with students in educational artscience labs, maturing through public exhibition in cultural artscience labs, and turning out socially innovative products in commercial artscience labs. Repeated experimentation, the exhibition and critique of prototypes, and team evolution give substance to the walls that guide ideas through this laboratory funnel. Partly, ideas succeed because a system of lab management monitors their vitality. But mainly ideas succeed because the passion of their creators pulls them along, acting like a gravitational force that propels ideas through the various stages of maturation as they move from personal, internal, or local critique to public, external, even global assessments of value.

Creative people and organizations thrive on dialog around creative ideas—and also occasionally fear it. While dialog can teach by pointing out propitious new directions, it can easily disappoint too, bringing on judgment without complete information. Presentation of an idea, as in a final class or through an art exhibition, generally does not promote very satisfying give-and-take. More typically it leads to various isolated data, like written audience feedback, test results, press reviews, and visitor or reader numbers, which the creator and lab do their best to understand, in the hope that they will set the idea on a new path.

At many points in our lives we all face this iterative process of idea development, and we do many things to diminish the risk of pain. This can mean simply failing to show up—by quitting a job, flying off in a plane, or just refusing to get out of bed. We may also speak in jargon, dress in certain distinctive ways, and work or play in specialized places, all out of a sense of self-protection. But what keeps our ideas in isolated environments eventually drains us of inspiration. If we're fortunate, we stumble upon an idea funnel of the kind I have been describing here. A child tries out a magic trick on his brother, then a parent, then a teacher, and finally a class. A choreographer tries out a new dance with a mate, later with skilled dancers, later still with a critic. Such funnel approaches help us develop ideas by trial and error and move them on to progressively less specialized environments while minimizing the pain that comes from the suggestion that we've simply got our ideas wrong.

It is in some way a core mission of laboratories that they provide mechanisms to help ideas mature and eventually reach audiences outside the lab. Recognizing the vulnerability of un-realized ideas when separated from the passion of their cre-ators, the most successful labs tend to move creators, along with their ideas, to new creative environments. This is as es-sential for the ideas as it is for those who come up with them. Creators are less original when they develop their fledgling ideas before the same small, predictably supportive groups of people. Guided in predictable ways, creators will do predict-able things. They surprise less.

Predictability also lowers the sense of collective risk that binds creators to a widening circle of collaborators with whom they innovate. New and critical collaborators come on board as ideas move through the idea funnel. The stakes go up, and this compels collaboration. Eventually, if we are not all taking a big risk together, our dependence on one another diminishes, we care less about what others do, and we cease to care all that much about our collective creative existence.

An idea funnel, then, keeps creative energy high by moving ideas as quickly as possible to new and more publicly engaged lab environments. Not all ideas make it through to broad cul-tural, commercial, or humanitarian engagement, nor should they. Each art and design experiment stands on its own—and it must, or else what is taking place is not really an experiment at all. In one guise an experiential learning program for stu-dents, in another an exhibition gallery for public dialog, in yet another a design and production studio, the artscience lab supports creators wherever they are in the idea funnel, from conception to realization of their ideas.

＝＝

What constitutes the investment value of an artscience lab? We measure value in scientific laboratories by the number and quality of publications and the public impact of innovations that emerge—and we can measure value in artscience labs in similar ways. The Wellcome Collection, which aims to bring its vast medical science compilation of objects into contemporary public discourse, receives around 330,000 visitors per year. The Science Gallery, which divides its mission between public outreach and service to the Trinity College community, receives around 156,000 visitors per year.

The first few years of Le Laboratoire, where the mission is to develop the innovation potential of experiments, has seen overall audience numbers grow as its art and design ideas have progressed from first-time exhibitions. Exhibitions in Paris attract around 15,000 visitors each year, while second- and third-time exhibitions in galleries and museums in Brussels, New York, Hong Kong, Tokyo, Bangkok, Basel, Graz, and Copenhagen have raised annual audience numbers to between 500,000 and 1,000,000. Purchasers of first-generation commercial or humanitarian-oriented products from Le Laboratoire were around 50,000 in 2009, two years after the opening of Le Laboratoire, and are estimated to be around 500,000 in 2010.

The labs described in this book mostly follow this innovation model, where value grows more apparent as ideas move along from conception to realization. Just as first publications of scientific advances share a creative process that can be readily extrapolated only by the specialized few, cultural exhibition in artscience labs brings ideas to the public first in intimate contexts and later in progressively less intimate contexts as the idea develops. This process of cultural incubation is, to return to our earlier analogy, akin to the creative process of our own lives.

By living we change the world around us to whatever degree. What comes from this change will be for the better, we hope, but we can't be sure, and experience suggests that some days will be better than others. We look for feedback, data, and reaction from those around us, to help us understand how we should live our lives, where we should go next with our ideas.

Let's say that I wake up one day and wish to speak to an old mentor with whom I haven't communicated in years. What do I do? It is the start of an idea translation process, a kind of experiment. Perhaps I hypothesize that my mentor will be reachable by email and not offended that I have approached him so informally after all these years. He was once like a father to me. I ask my wife what she thinks. Send the email, she says. So I do. There is no response for days. He isn't going to respond, I decide. An experiment is over, I've learned something, and it's not entirely clear what I should do next. I call a friend, who knows my mentor and knows me. I ask what has happened to my mentor, what state of mind he is in, and get the advice to stop by his office next time I'm in Cambridge. My mentor would prefer to see me, not read an email from me or hear my voice over the phone. So I do this. I walk into his office two months later and find him sitting behind his desk. We're both surprised by this situation. Neither of us could have imagined it, could have known that he would find me standing in front of him, or I would find him sitting there. We're both moved. I did the right thing in part because I shared my idea as I developed it. I learned, others learned, and something good came of it.

Looking back on an idea translation process, we often remember the pivotal moments (such as the phone call to my friend) that separate one hypothesis-driven experiment from the next. We stare at what seems to be a blank page. These can be times of confusion, uncertainty, even trauma. We wonder what to do next. Sometimes we wonder for days, even months or years. Our previous experiment did not exactly fulfill our dream, perhaps was not even intended to fulfill the dream but was designed, rather, to get us closer to fulfillment. Now, given what we learned from the previous experiment, we must decide what to do next. We don't quite know. We deliberate over this blank page, perhaps for a very long time, and eventually, often provoked by someone to whom we offer privileged insight into our idea—to whom we "exhibit" our idea in a way— we imagine something new. We analyze this idea, frame a new experiment, and execute it.

The artscience process involved in carrying us from the contemplation of the unknown through the admission of uncertainty to the arrival of a new idea is the core process of a creative life. It fuses the aesthetic with the analytical, especially at these blank-page moments that we sometimes look back on as the moments of crisis.

This book describes a special kind of lab and cultural incubation process that make this idea translation happen more readily and with greater chance for valuable impact. It shows how a lab led by art and design experimentation can help creators realize their dreams. It also shows how the artscience lab partners with and benefits large educational, cultural, humanitarian, and commercial organizations, ranging from universities, such as Harvard and Trinity College, Dublin, to museums, such as the Louvre, and humanitarian organizations, such as the Bill & Melinda Gates Foundation.

In an age when society and culture are rapidly evolving, large institutions need to adapt if they are to respond positively to the needs and opportunities of a changing world. Innovation is paramount. But institutions where specialization of information and function discourage innovators and dampen dreams (if merely with the purpose of mitigating risk) are not well suited to innovative practice. They innovate best through association with innovation catalysts. And this is what artscience labs can be. Artscience labs provide interdisciplinary conditions and expressive environments that benefit creators while fostering productive relationships with established institutions, to bring about beneficial, sustainable change.

≡2≡
ARTSCIENCE LAB

A few years ago, Sian Ede announced in her wise book *Art & Science* a phenomenon that was turning up in the mathematical, biological, physical, chemical, applied, and social sciences. Artists were exploring the meaning of scientific frontiers— sometimes working with scientists, sometimes working alone, and sometimes even looking like scientists themselves. More recently, Stephen Wilson's book *Art & Science Now*, with its catalog of contemporary artwork inspired by science and technology, has illustrated more visually where artists in the field have been going.

What these artists do, and sometimes say, is enlivening the debate around creative process, even while it is inevitably erasing some of the convenient, simplifying distinctions we have long made between art and science, artists and scientists, and the exclusive roles and objectives of each.

In the next pages I describe how a new kind of artscience lab can aim this kind of artist experimentation in the sciences toward surprising innovation in society through a creative spirit forged in the process of its own creation. The innovation models of Google, Ideo, MIT Media Lab, and the Ars Electronica Futurelab, to which I will refer in the remainder of this book, all emerged in this way.

Google, for example, spun out of an academic lab at Stanford University where both Larry Page and Sergey Brin worked on the Stanford Digital Library Project. The company emerged through an iterative process of doctoral research and public feedback that took in ideas and data as rapidly as the Internet could deliver them. Since neither Page nor Brin urgently needed revenues, time mattered more than money. And because they were not alone in the race for a good search engine, innovation mattered more than perfection, and they valued ideas based on the amount of hard data those ideas generated, no matter what their source might be. The Google model—innovate first, perfect later, share everything, pursue brilliance, dare to dream, and believe that data and money follow the idea—came about from the way Google itself came about.

The same was true of the Media Lab at the Massachusetts Institute of Technology. Its model formed around founder Nicholas Negroponte, a charismatic architect in MIT's School of Architecture and Planning. To pioneer research and spark development in digital media, the Media Lab adopted a work philosophy suited to a designer: open studio spaces, research teams, few intellectual property hurdles, and a spirit of "demo or die" (which designer John Maeda eventually rephrased as "imagine and realize"). The MIT Media Lab's model for innovation—passion, experimentation, interdisciplinary collaboration, rapid prototyping, partnerships with industry, and solid grounding in practical problems—grew out of the lab's own creation story.

Ideo began as the brain child of David Kelley, who completed a graduate design program in Stanford's School of Engineering and then started a small firm in Palo Alto in 1978. Kelley had met Steve Jobs at Stanford and by 1983 helped Apple design the first commercially available computer mouse. As the company grew, it merged in 1991 with two other innovative design firms, ID Two and Matrix Product Design, to create Ideo Product Development (later simply Ideo). The company's innovation model, captured in Tom Kelley's *The Art of Innovation*, and, more recently, the book *Change by Design* written by Ideo CEO Tim Brown, came about in the fast-paced atmosphere of Silicon Valley in the 1980s and 1990s. Its basic premises—understand, observe, visualize, implement, and evaluate—are those of the engineer, while its reliance on teams, a flat hierarchy, brainstorming, and rapid prototyping reflect the philosophy of the practicing industrial designer.

The Ars Electronica Futurelab innovation model has commonalities with the MIT Media Lab model, in that both are focused on media arts, though Futurelab puts greater emphasis on breaking down barriers between artists and scientists. This mission reflects its origins in the success of the Ars Electronica Festival, the preeminent international Cyber Arts festival in Linz, Austria. Interdisciplinary teams with artists, computer scientists, physicists, game developers, sociologists, and others all work together through what Ars Electronica calls "shared creativity" in a large lab-atelier environment that evolves with the nature of the project.

Le Laboratoire in Paris and its network of artscience labs, whose startup I describe here, share many features with these innovation organizations. Interdisciplinary collaboration, rapid prototyping, exhibition or demonstration, and translation of ideas into products or processes with beneficial social impact all play a fundamental role in our concept of "idea translation." Perhaps the most distinctive characteristic of artscience labs is that education, cultural exhibition, and production are all simultaneously core values. The diversity of these values helps give particularly far-fetched ideas time and place to mature as they make their way through the idea funnel of the artscience lab network and outward to the larger world.

⟹⟸

Startups are, of course, risky adventures. A startup may not become what its initiators imagined: their dreams may fail, or, just as bad, the startup's particular innovation may not be understood, appreciated, or accepted, even by those who need it most. Failure by either route can mean the loss of time, money, even friends. To improve their chances of success, startups generally bring together two or more people who share a dream and its associated risks, and who hope that what may happen tomorrow is worth the trouble today.

Risk sharing in the pursuit of a collective dream is not just necessary, it accounts for some of the great unspoken pleasures of creation. Risk-taking is a collective bet on the reliability of every person involved in the venture. At a time when the technologies of modern civilization seem to diminish our need for others, this sudden clarity of mutual dependence ("I can't realize my dream without you") can be surprising, refreshing, often moving, and can work as a kind of guide to creative collaboration.

Recently, my eight-year-old son Raphael, having meditated for a couple hours on the film *Indiana Jones and the Last Crusade*, spontaneously assumed the role of Indiana Jones, while his younger brother Thierry took on the role of Jones's supportive friend Sallah. They would not have performed as they did alone—for one thing, performing alone risks drawing smirks from a brother. It is also easier to improvise a twenty-minute skit with a sibling than to make it up all on your own. A quick suggestion by one, a fast refusal by the other, thirty seconds of negotiation and compromise, and within a minute Raphael and Thierry were productively collaborating. This helped Raphael overcome his innate shyness and fulfill momentary ambitions of adventure and glory. Thierry, who hates to follow the lead of his two brothers and occasionally gets into trouble because of it, got to be the ham while showing how useful he could be to Raphael, for once.

My sons assumed the risk of performing for rewards that mattered to each individually. But there was a larger interest, too. They reenacted a scene from the movie—a collage of scenes, actually—which managed to make their parents smile and applaud. By pursuing their vision and sharing its risk, Raphael and Thierry created something whose value exceeded the combined value of what each could have experienced on his own. It would be wrong to see Raphael as the guiding force of their collaboration, just as it would be wrong to interpret Thierry's role that way. And it would be especially wrong to believe that before their brief conversation either of them had the slightest idea of what they were about to do. They just figured it out, developed their idea on the spot—passionately and collaboratively—and had a ball.

We see creative banding like this often enough, but for the most part we live and work within specialized organizations, among people whose education, competencies, and aspirations are much like our own. Institutions like this tend to take a different approach to collaboration, seeking above all else to minimize the risk of failure. As a consequence of this tendency to stick with those whose work we understand, shared risk-taking remains one of the least understood and least appreciated factors in creative collaboration.

At several critical moments in the conception of the art-science lab network created around Le Laboratoire, risk-taking built cohesion while helping forge connections among the labs of the network and with other organizations. This web allowed innovative ideas, each with its own associated risks, to transfer from one organization to the next more readily than would have otherwise occurred. Relative to other innovation models, nurturing and developing particularly risky ideas is the essential added value of artscience labs.

=≡≡=

Six months before the opening of Le Laboratoire seemed the right time to visit Hans Ulrich Obrist, director of the Serpentine Gallery in London. A brilliant synthetic thinker, Hans had collaborated in Antwerp with Barbara Vanderlinden in 1999 to curate a famous citywide exhibition, *Laboratorium,* which explored connections between art and science.

"Famous, yes," said Hans, "which I find ironic since almost nobody saw it. You have to be careful getting too near contemporary science. And patient. *Laboratorium* is mythic because it was right. One of the most interesting places to be as an artist is in the mystery world of science. But bringing art to science gives you a good chance of having neither." Obrist pointed to papers pinned to a wall behind his chair. For some time he'd been asking artists to send along their "formula for life." A few scientists had contributed as well. James Watson had drawn a double helix.

In *Laboratorium,* Obrist and Vanderlinden invited artists and scientists to open up their labs, philosophers and historians to explore the meaning of a lab, and the public to consider how to define a lab, today and yesterday. There were experiments, like those curated by the French historian of science and philosopher Bruno Latour, but *Laboratorium,* even as an exhibition of contemporary art, had not sought to create art at the frontiers of scientific knowledge. Obrist's exhibition, rather, invited the public into an exploration of the lab as a "knowledge-generation institution," in the words of Peter Galison, who contributed to the exhibition catalog as a historian of science. Seeing the lab this way meant that ideas did not flow—come into and leave with greater likelihood to live on, as was the case, for example, in Andy Warhol's 1960s New York studio The Factory. Obrist raised the subject of The Factory in our conversation. Was it a model for Le Laboratoire?

The Factory had provocatively merged art creation with industrial silk-screen production, which in a sense paralleled the rapid transfer from creation to production that was then taking place in the science lab. In Warhol's Factory, High Art looked a little like High Science. *Laboratorium* invited you into the lab to observe and learn. Nobody invited you into The Factory; or—if you were invited—it was at your own risk. The Factory was a place of privileged creation and production, mostly for Warhol himself, but also for a few walk-ons and recruits, such as the singer and songwriter Lou Reed, discovered by Warhol during a performance in Greenwich Village. An invitation into The Factory became the creative stimulus for Reed's hit "Take a Walk on the Wild Side."

The Factory seemed indeed closer to the vision of Le Laboratoire than did *Laboratorium*. At the time of the Serpentine visit, Le Laboratoire was to be a lab where artists and designers performed experiments at the frontiers of science. Experiments would not be purely illustrative, and not limited or transformed by the need for a particular result. Experiments also would not require residencies and might happen inside or outside the lab. Exhibition would be a kind of initial peer review open to the public, though it was likely that exhibits would have narrow public appeal. But while it was hard to predict who would wish to see an exhibition, what seemed most important was that the experiments intrigue artists, for the adventure, and scientists, for the chance to explore the edges of knowledge. Le Laboratoire's founders also hoped that institutional relationships would be easy to forge. On one side, we expected that students from Harvard University and other schools would come to the lab to develop ideas that originated in experiential learning programs. On the other side, we hoped that for-profit and nonprofit partners would work with the lab to carry these ideas forward into commercial or humanitarian ventures. If all this happened properly, the finances of Le Laboratoire would balance in the way that finances typically balanced in any successful science lab.

This business model, however, was based on a track record we did not have in those early days. Le Laboratoire came into existence predicated on an assumption that its founders' intentions would be generally understood—an assumption that quickly proved wrong, as many of our early assumptions did. Yet it was through adjusting, as any creator adjusts before arriving at a finished work, that the lab formed its special innovative spirit: to learn by creating something that places you at risk, and then to create in response to what is learned.

As David Pye points out in his classic *The Nature and Art of Workmanship,* creation typically comes out of "workmanship at risk." We learn by making mistakes—or, in Pye's craftsmanship context, by making adjustments as we learn along the way. Because we take personal risks in what we do, we learn more quickly, and deeply, than we might learn had we nothing personal at stake. The process of idea translation—where each node, or idea turning point, in the iterative process of idea development becomes the trampoline for a new idea translation—blinds us to the potential for failure. If we don't notice failure, it is because each failed idea becomes the springboard for a new and better one. All of this, familiar to anyone who has started a new company, whether in the theater business or in the biotech business, builds team cohesion around the creative process and makes members of the team believe that the creative process itself, unpredictable as it is, will eventually lead to success.

Recognizing in its early days the need to attach a more concrete, if only more quickly comprehensible, value proposition to its creative work, the Laboratoire team considered relating its art and design experimentation to the innovative learning potential among low-income urban youth. Artists we worked with either appreciated the idea of involving urban youth in what and how they created, as Fabrice Hyber did, or they were already working actively with urban youth, as was the double Michelin-starred French chef Thierry Marx. For several years, in Boston and Paris, my wife and I had worked with teens through the arts and found them especially motivated by the rewards of creation and full of fresh things to say. This vision of making an impact on young lives inspired us, and we hoped it would inspire others.

But in the fall of 2005, as leaders of Le Laboratoire tried to rally investors, unrest broke out in Paris. Several of the suburbs—where unemployment ran above 25 percent, offering no future for the kids, and where the war in Iraq was interpreted as a wider war on Islamic culture—exploded with acts of vandalism and violence. Youth from the communities where we worked began burning cars—thousands a day. Our neighbors, meanwhile, uncovered evidence on the Internet that in previous years the lab's founders had thrown hip-hop parties inside the Eiffel Tower and invited what seemed to be the same suburban youths who were now burning cars. This connection was seen as a warning of what their own neighborhood might become. Le Laboratoire's social intentions meant little to these neighbors, while the risks seemed all too real. So, in classic and predictable fashion, the neighbors fought the lab in building-management meetings and finally approached the neighborhood prefect, who acceded to their wishes by refusing to grant a building permit.

A year away from the opening, Le Laboratoire remained in need of a near-term value proposition that would motivate investment in the project, not to mention allow construction on the site to continue. The lab's leaders, including Caroline

Naphegyi, formerly curator of citywide exhibitions in the northern French town of Lille, and Olivier Borgeaud, formerly cultural attaché of the French Embassy in London and Berlin, looked to philanthropic support. Perhaps philanthropists would see the cultural value of what we wished to do. But our bids for sponsorship from banks and large utility companies were unrealistic, and in the spring of 2007 they too fell through. Traditional sponsors simply could not understand in those early days exactly what the lab wanted to accomplish, or perhaps even why. Six months from opening day, with the building permit at last issued, there was still no funding in sight.

Stressful as all this was, the early days were wonderfully exhilarating. If potential supporters could not quite understand our carefully chosen words, this meant only that we needed to produce results more quickly than we had first realized. There was an exciting sense of urgency to what we did. The amplified risk made members of the team need one another more than ever before, and they became better collaborators. Le Laboratoire itself was an experiment with a very uncertain outcome, and pursuit of this shared dream proved as exciting as the thought of its realization.

In late August 2007 the first article on Le Laboratoire was to appear in the leading monthly culture publication in France, *Beaux Arts Magazine.* Would it call attention to the breathtaking economic uncertainty of the project? Had that happened, Le Laboratoire might have closed down even before it opened. Instead, in a gesture of blanket support, considering where things stood, the article heaped praise on the lab initiative. On reading the article, we quickly sent a text message to the magazine's editor, Fabrice Bousteau, expressing thanks, to which he immediately responded with the generous quip, "No, thank you."

≡≡

The first year of Le Laboratoire started with the theme of intelligence, which was broadly relevant to contemporary scientific research yet still ambiguous and therefore inviting to artistic speculation. A painting in Fabrice Hyber's studio showing apples falling to the ground to become grapes prompted him to ask if Le Laboratoire could make something like this happen. His question was confusing. What did the theme of intelligence have to do with a painting of an odd tree? The question went unanswered until it became a different question entirely: what if Fabrice could see his apples as stem cells and his grapes as neurons? This might bring him to a frontier of scientific research related to intelligence. Robert Langer and his team at MIT had for some years been fashioning neurons from stem cells and implanting them in animals with the hope of eventually correcting spinal cord injuries, and this work, viewed from the perspective of Le Laboratoire, related in a few ways to what intelligence was and was not. Fabrice liked the idea. He traveled to Cambridge, where he met with the Langer lab. Fabrice then had the idea of creating an installation that would give a visitor the sense of what it meant to be a stem cell becoming a neuron. Musing that if you could fall through an hourglass you might share some of the experience of cellular division, Fabrice and the team at Le Laboratoire made giant inflatable hourglasses and more than thirty paintings and sculptures, including a neuronal axon-like form comprised of four thousand pieces of strawberry-flavored bubble gum heated up and stretched down to the cement floor from an iron beam on the ceiling.

Visitors to this first exhibition expressed amazement and at times frustration. There were no guides, little explanation of anything. Even Mathieu Lehanneur's more comprehensible exhibition, which displayed a new plant-based filter (Bel-Air) that later became one of Le Laboratoire's first commercial products (renamed Andrea), seemed to mystify many. Did it work? We didn't know yet. Could you purchase it? Perhaps one day.

Le Laboratoire's second exhibition raised even more questions. Led by *Time* magazine photographer James Nachtwey in collaboration with medical scientist Anne Goldfeld, the photographic project depicted the pain and compassion of sufferers and healers in the war against AIDS and tuberculosis and posed moral questions. From bubble gum attached to the ceiling to startling images of the dying and the dead was a thematic distance too great for many. Attendance dropped. But the exhibition proved effective in what it aimed to share, a process of collaboration between artists and scientists in a war against deadly disease. The artist Asa Madar shot poignant footage of Anne and other medical scientists, and the exhibition served as the setting for an international conference in February 2008, attended by about a hundred scientists and healthcare workers from around the world, with funding from the Bill & Melinda Gates Foundation. Leaders of the foundation asked to bring the exhibit to Bangkok the following autumn for the annual Gates Grand Challenge Meeting, the first exhibition that traveled from Le Laboratoire.

By the spring of 2008 rumors swirled—at parties, in cafés and restaurants, during chance encounters on the street—about the limited future of Le Laboratoire. The lab had practically no external support and could not continue without it. But simmering alongside all this financial distress was a new source of hope. Our third exhibition in the spring of 2008 featured French chef Thierry Marx. The intersection of culinary art and science was by then a strong movement pioneered by the French chemist Hervé This, renowned chefs like Ferran Adrià at Spain's El Bulli, and Heston Blumenthal at Britain's The Fat Duck. Thierry's own restaurant was located at Chateau Cordeillan-Bages in Pauillac outside Bordeaux. He had visited the lab's construction site months before the opening and eventually worked with the team to hatch an idea for an experiment in collaboration with the French colloid scientist Jérôme Bibette of the Ecole Supérieure de Physique Chimie Industrielle. Jérôme and Thierry wanted to develop "spherification," an improved method for encapsulating flavors. Jérôme did it in a way that eliminated the thick membranes of lecithin that had until then been used to make tiny eggs of flavor. But he was not able to complete his work in time to encapsulate salmon, lemon, or tomato with the quality of taste you anticipated from the chef of Cordeillan-Bages.

The exhibition opened with the largest wave of press coverage Le Laboratoire had seen until then. But we had to close down just after the opening for three weeks while Jérôme and Thierry worked to get the process—and the catering staff—

to some minimum level of effectiveness. The exhibition re-opened, and by the time it ended three months later it had drawn the most consistently engaged audiences in the lab's short history—fifty to a hundred visitors a day, each spending an hour or more participating in a culinary art and colloidal science experiment.

Thierry's experiment formed the background to another idea, which we called Le Whif: breathable chocolate and, later, other kinds of food. This idea was a twist on an approach I had used in my scientific research to prepare vaccines for delivery to the lungs. It brought culinary art to a frontier of medical science with an intention to learn from the encounter. Exhibition visitors had fun with Le Whif, and we guessed there might be an opportunity, a chance for a second commercial innovation (with our emerging air filter Andrea), though no potential investment partner expressed any interest. To pursue development of Le Whif and Andrea, Le Laboratoire needed to start a companion artscience lab dedicated to projects with commercial or humanitarian potential.

José Sanchez, one of Thierry's business partners, became president of this new venture, which we called LaboGroup. He arranged for a first prototype of Le Whif to be manufactured in China (where he started work on Andrea, too) and over the next twelve months began recruiting other talented people to his team. The business model of LaboGroup added to the risk of the still-unproven business model of Le Laboratoire. There were no good data to quantify the commercial opportunity Le Whif or Andrea offered. And what we initially made and presented as Le Whif and Andrea was far from product-worthy. José and his team had caught the spirit and placed a real bet on the model of Le Laboratoire. There were now two labs that experimented in the arts and sciences with the intention of translating surprising ideas into commercial and cultural practice.

We opened the second year with the artistic outcome of an experiment between Japanese digital composer Ryoji Ikeda and Harvard University number theorist Benedict Gross. Ryoji had come to Harvard to meet Dick in January 2008. The meeting, on a Friday evening in a conference room near my office in Pierce Hall, almost immediately turned into a computer-generated demonstration of Ryoji's unforgettable and astonishing *datamatics* performance. After some discussion, Dick raised the idea that Ryoji might help mathematicians visualize Cantor's Set, a collection of numbers obtained by extracting a third of all real numbers between zero and one, and then a third of the remaining numbers, and repeating the operation an infinite number of times. Mathematicians couldn't visualize in any satisfying way what remained. Logic argued that nothing should remain, but in fact the set was infinite, dots scattered widely within an infinite space. Over the weekend Dick gave Ryoji a short course on the subject, and plenty more discussion followed. Nine months later, Ryoji finished his $V \neq L$ exhibition, a composition without sound, as he put it—two five-meter images that each contained seven million digits spread out over a black carpeted floor with light slicing in from a second installation where you walked through a corridor of white light.

By then Le Whif was selling modestly in a new LaboShop that had opened inside the growing Laboratoire space in Paris, Andrea was in early commercial production, and a third artscience lab had opened up: the Idea Translation Lab at Cloud Place in Boston, run by the Cloud Foundation, which brought inner-city teens into the artscience lab network to pursue innovation in education—a goal of ours from the start.

In the winter of 2009 Le Laboratoire's exhibit explored fear and terror through an experiment by the Indian artist Shilpa Gupta in collaboration with the neuroscientist and psychologist Mahzarin Banaji. This exhibit generated a stream of favorable press and accompanied the startup of two other artscience labs: another educational lab at Harvard University, the origins of which predated Le Laboratoire, and humanitarian innovation labs in Pretoria and Cape Town, a collaboration between Harvard University, the nonprofit Medicine in Need (MEND) I had started in the early 2000s, and the University of Pretoria.

By the winter of 2009 it was clear that all these labs—the three idea translation labs in the United States, France, and South Africa, and Le Laboratoire and LaboGroup in France—formed a loosely knit interdependent organization. The various artscience lab teams were in regular weekly discussion, within and across organizations, betting on the competence of the network not to drop an idea that had developed in, perhaps, Le Laboratoire or to develop an idea that, perhaps, Le Laboratoire would eventually need. The labs faced communication obstacles working across three continents, time zones, and organizational missions. But by passing ideas around among themselves, all the labs shared the risks. Communication obstacles fell as mutual dependence and trust grew.

≡≡

Le Laboratoire's first trickle of revenue began in late 2009 from sales of its first two products. The lab had oriented its experiments toward four paths of development—humanitarian, cultural, educational, and commercial—before one, the commercial, began to pay off. Then almost simultaneously the others did, too.

As it turned out, Le Laboratoire's cultural incubation model itself became the most enduring product of its startup network. The exhibit space in Paris was at the heart of our cultural incubation approach to innovation, the place where the public engaged with ideas coming from the lab and where the press opined about what these ideas meant in cultural terms. No experiment happened at Le Laboratoire that did not also offer an opportunity to learn through the process. From the exhibition of Fabrice Hyber to the museum guide technology MuseTrek (our third commercial venture), team members experimented, learned, and "published" the lab's results through public exhibition. These exhibitions exerted a kind of pull on students from Harvard University and Cloud Place, who more passionately pursued their dreams in the idea translation labs, knowing there could be a chance to develop them further in Paris. Meanwhile, Laboratoire exhibitions pushed ideas out the other end of the funnel through LaboGroup, as commercial, cultural, and humanitarian outcomes.

Exhibition at Le Laboratoire became a showcase of ideas, a chance for fresh feedback, and the starting point for a public relationship with the innovative process. The press had a chance to participate in the evolution of an idea without demanding the clarity of a mature product. Indeed, the immaturity of the idea, like the whispers and wails of an infant, became its strength, an invitation to participate in idea growth. Ideas that traveled through this idea funnel grew more surprising and innovative precisely because at the outset no one felt that an idea absolutely had to lead to some kind of beneficial change. Ideas were expected to grow up at their own pace.

The network of artscience labs developed an innovation model founded on its four pillars of cultural, commercial, humanitarian, and educational values.

Cultural Value

Early on it would have been difficult to assign any value other than this to our venture. Still, most who read or heard something about Le Laboratoire's art experiments in Paris and elsewhere could not really say what the lab was doing. Audiences numbered consistently around a few thousand per exhibition, a hundred or so visitors on a good day. But word got out. Le Laboratoire was a surprise. Almost nothing in the cultural experience of Paris prepared visitors for the lab's exhibitions. Early on, few journalists could say even what Le Laboratoire was. They searched for words. Le Laboratoire seemed out of touch, perhaps elitist. Certain visitors seemed intrigued, but perhaps this was simply the pleasure of surprise at what the lab was doing in comparison with the predictability of whatever else was available on the cultural scene at that moment. This was not much of a selling point.

But there was something else. Those few who appreciated what was happening at Le Laboratoire came with special seriousness. Students at our first exhibitions sat for hours on the floor, meditating. Others leaned against a pillar and took notes. Many hung around the exhibition space discussing science and art with student guides. The issues explored by the lab in its second year started to attract the attention of large cultural institutions in Paris. We began to explore the value in partnering and sharing production and exhibition costs and mitigating risks on both sides. The Shilpa Gupta exhibit represented our first experimental collaboration with the Louvre. The following exhibition led to a collaboration with the Musée des Arts Décoratifs, and by the winter of 2009 collaborations were under discussion with the Centre Pompidou. Le Laboratoire sold its first produced work of art to the Louisiana Museum of Modern Art outside Copenhagen in the fall of 2009, and catalogs of the Ryoji Ikeda exhibition traveled to The Laboratory at Harvard when it opened in November 2009. From that date, Le Laboratoire managed to find institutional partners for each of its exhibitions.

Commercial Value

Throughout its first year, Le Laboratoire operated with a single door accessible from the street. Seventy percent of the lab's façade remained boarded up as team members deliberated over what to do with it. Looking like a sad blemish on the cheek of an adolescent, two hundred square meters of street-side commercial space remained shabbily unused for months after Le Laboratoire opened. The safest thing would have been to rent it out. There were conversations about this with the Galeries Lafayette, the New York-based Material Connexions, and others, but these established businesses found little to interest them.

Then, midway through the Thierry Marx experiment, in the spring of 2008, while burning through cash at an alarming rate, the team thought up the notion of creating a space for brainstorming (a few team members still needed offices), a prototype design store (Le Laboratoire by that time had things to sell, though the marketing would require a special touch), and a FoodLab (Thierry needed dedicated space and equipment to continue his culinary art and science experimentation). Thinking it through over the summer months, we saw how the first space, with windows on the street and a giant white board, would be the one quiet spot at 4 rue du Bouloi— a place to contemplate in full view of the neighborhood. The second space would be a kind of commercial test lab, where experimental designs could gradually grow up and get ready for the real world. The third would be a culinary lab. All this was worked out over lunches with the designer, Mathieu Lehanneur, by then a close collaborator.

Construction of these new spaces was completed around the time the world economy sank in October 2008. The blight of boarded up windows disappeared and made way for a unique space whose street-side weirdness adequately reflected our purposes. Le Laboratoire now had three times as much façade as before. The brainstorming space, which doubled as my office, drew street-side attention with its white-board grotto and big beanbag, where I sometimes sat to work with curtains drawn back. The LaboShop opened to sell a cup of coffee and a whiff of chocolate—our prototype food innovation was taking its first commercial baby steps. The FoodLab with chef Marx and his apprentice Thierry Martin offered occasional tasting lunches and dinners and collaborated on the fledgling aerosol cuisine business. By October 2009 revenues from sales of Le Whif and Andrea started to grow. These products went initially on sale at the LaboShop and also through stores around France, including Monceau Fleurs, Nature et Découvertes, Galeries Lafayette, BHV, and, soon, internationally, through Internet marketers such as Amazon, Hammacher Schlemmer, and FrontGate, and visible stores like Dylan's Candy Bar in New York City, the MoMA design store, and House of Fraser in England.

Humanitarian Value

Medicine in Need, a nongovernmental organization based in South Africa, the United States, and France, had come about to combat diseases of poverty by developing novel technologies and strategies for delivering drugs and vaccines. MEND, which grew through the 2000s with funding directed to my Harvard lab from the Bill & Melinda Gates Foundation, introduced us to an altruistic realm of science where talented healthcare workers, clinical investigators, and administrators had turned their efforts since the late 1990s. The idea that

Laboratoire experimentation might spark innovative thinking in the development of medicines with enormous humanitarian value but little commercial promise led to the experiment with James Nachtwey and, a year later, with Shilpa Gupta. Both produced artworks relevant to problems of global health, Jim purposefully aimed at HIV-AIDS and tuberculosis and Shilpa incidentally aimed at the growing crisis of mental health in the developing world.

Medicine in Need made these exhibitions the setting for scientific conferences. Under the guidance of the CEO Andrew Schiermeier and Bernard Fourie, head of the African organization, MEND gathered leaders from the United States, Asia, Africa, and Europe at Le Laboratoire. Then the Bill & Melinda Gates Foundation invited the Nachtwey exhibit to Bangkok for the annual meeting of the Grand Challenge Program, creating an unforgettable atmosphere crystallized by Jim's moving address in the plenary session. The World Health Organization held a meeting at Le Laboratoire during the Gupta exhibition.

Around this time, the lab began to develop a novel approach to transporting water in arid low-income parts of the world. This idea emerged out of a Harvard class and grew through the creative efforts of François Azambourg and Mathieu Lehanneur, its designers. Mathieu worked with us to create what we called the Pumpkin, while François helped us design new ways of transporting drinkable liquids that related to the form and function of biological cells. These products went on exhibit at Le Laboratoire in the fall of 2010, when Medicine in Need presented a business plan with the Labo-

Group to bring the Pumpkin to the developing world, driven by sales of the product in the developed world.

Educational Value

Each new experiment carried Le Laboratoire to edges of scientific knowledge, providing opportunities for learning by raising questions that mattered to groundbreaking artists, designers, and scientists alike. As described in my book *Artscience*, the original idea for Le Laboratoire had emerged in my work at Harvard University and had grown with the passion of students there. In Fabrice Hyber's experiment, Harvard students collected data related to neuronal engineering and material science; in Mathieu Lehanneur's experiment, they did filtration research, learning how and why plants absorb toxic gases that limit their efficiency; in James Nachtwey's experiment, they animated an international conference on global health delivery technology; in Thierry Marx's experiment, they formed a team to brainstorm the creation of Le Whif; in Ryoji Ikeda's experiment, they explored the art and science of the theory of numbers; and in Shilpa Gupta's experiment, they investigated the origins of fear and participated in the interactive experiment at the Louvre around MuseTrek.

The educational arm of Le Laboratoire was the idea translation lab, which provided mentorship and guidance for students as they developed their ideas, and then funded summer idea-translation experiences for many of them around the world. Starting in 2008, students from Harvard, the Boston

Public Schools, Trinity College in Dublin, Strate College in Paris, and other schools came to Paris each summer for a ten-day Innovation Workshop, where artists and scientists helped them develop ideas that ranged from a project to light London in some original way for the 2012 Olympic Games to composing music by processing signals emitted from solar energy. This program grew, and in the late fall of 2009 led to the opening of a new art and science cultural center, The Laboratory at Harvard, in Cambridge, Massachusetts.

≡≡

Science seems to advance by an occasional boom rising over a chorus of mysterious whispers. It is much easier to say what scientists have discovered, or might discover, than to explain how or why exactly they discover anything. We can read about scientific results in the peer-reviewed literature and peruse popular commentary as to what it all means. But what's harder to discern is why science advances more fruitfully in some labs than in others, even when members of these labs have equivalent education, professional training, funding and other resources, and access to the peer-reviewed literature.

In recent years, several artscience labs have appeared around the world to explore this question of creative process and to invite the public to participate in answering it. Some of these labs now form loose networks where art and design ideas at the frontiers of science can propagate in full public view, permitting a deep look into the creative process through the prism of culture. With an entrepreneurial and risk-taking spirit formed by the very conditions under which they got their start, these labs have emerged as catalysts for innovation in a broad sense. They provide opportunities to reinvigorate the peculiar innovation model of the contemporary science lab, while encouraging artists and scientists to expand the notion of risk-taking and creative banding that is at the heart of contemporary media-innovation models.

≈3≈

EDUCATION

The educational artscience lab engages high school and university students in experiential learning through the pursuit of innovative dreams. We call this organization the "idea translation lab." As the broadest entry-point to the idea funnel, the idea translation lab helps students "learn to learn" in real-world settings while pursuing dreams at frontiers of knowledge. Student projects begin as "seed ideas" proposed by artists, designers, scientists, and entrepreneurs and evolve from there through student initiative and creativity into collaborative ventures in art or design, and often some form of lasting implementation in society.

Learning by idea translation may appear to be unorthodox, but in a way it is the oldest form of learning we know. From the crib to the living room, from the living room to the playground, from the playground to our first class: when we were children, changing the rules, objectives, and setting often catalyzed learning. Having to adapt to new circumstances made us—sometimes excitingly, sometimes cruelly—sensitive to the enormity of what we didn't know. But as we grow familiar with the worlds in which we work and play, and as we specialize, this learning catalyst of youth wears away. We lose the habit of springing forward, not quite prepared. This may mean fewer unpleasant surprises, less confusion, but it also probably means diminished curiosity and personal growth.

Deliberately putting themselves in strange environments, as a stratagem to recover the wonder of learning, is an old trick of savvy creators. Research scientists—among the most highly trained and narrowly focused creators on the planet—have known for a very long time that they are more likely to make discoveries by consciously bucking specialization and romping outside their established fields of expertise, even when this negatively effects their publication rate in the peer-reviewed journals where careers are often made and by which success is usually measured. The learning that happens when scientists do this tends to be unlike the learning by which they specialized in the first place.

When a mathematician aims to develop an original idea in a fresh domain of knowledge, say cellular trafficking, she obviously must learn a great deal about the fields of cellular and molecular biology. But there is no need to learn these fields in the way that a student intent on gaining a degree in biology must learn them. The mathematician does not need to know everything about the cell or biochemistry, any more than a shopper for a pair of shoes needs to know everything about the business of shoe manufacture and retail to find the perfect shoe. The shopper learns enough to translate an idea into reality—to purchase a sturdy or stylish pair of shoes. The same is true of the mathematician looking to realize an innovative dream.

Scientific breakthroughs drive learning, and traditional schooling drives learning, but they do so in contrary ways. One starts from the particular; the other from the general. The first promotes surprise, discovery, and often uncertainty as to whether the discovery is even "true"; the second promotes repetition—discovered knowledge—and aims to diminish uncertainty. The former is like the voracious learning process of the preschool child; the latter is the disciplined learning process of the specialized adult.

Learning in an artscience lab is in this sense like child's play —it happens best when it is beside the point. Learning in environments where sometimes intuition matters more to learning than data and at other times data matter more to learning than intuition makes us, otherwise busy chasing our idea, wary of specialization. And this proves catalytic to creativity.

The present chapter describes this particular type of artscience lab, of which there are examples today in Boston, Cambridge, Paris, Oklahoma City, Singapore, and Dhahran. I especially draw on the experiences of two talented, passionate young leaders who made the first two idea translation labs succeed, Hugo Van Vuuren and Carrie Fitzsimmons.

When I first met Hugo in the fall of 2007 he was beginning his senior year at Harvard. An economics concentrator with a flare for new Internet ideas, he might not have noticed my idea translation course in the School of Engineering and Applied Sciences had he not heard or read somewhere of the nonprofit organization we'd started in Pretoria, his hometown. Curious about it, he came to see me a couple of weeks before class to learn more.

The idea translation class is open to all concentrators, from physics to literature. In the first week they form diverse groups of three to five students. In class they hear of far-reaching, vaguely formulated ideas, often put together with innovators outside the class, which may have purely cultural, commercial, or humanitarian relevance. For several weeks students mull over these ideas, each group honing in on an idea they feel passionate about—typically quite different from the seed idea they began with. This new idea, whatever it is, will aim at some major need or opportunity, with a path to realization that is so hypothetical it remains impossible to define beyond the first experimental steps.

Once a group announces an idea as its own, members of the group typically begin to doubt that it is actually achievable, the first step in the inquiry process. The job of instructors is to fan that doubt by making sure students realize how complex their problem actually is, and yet help them progressively overcome disbelief by expanding their learning and by imagining experiments that will clarify matters. For several weeks students develop a plan to translate their idea into a more advanced stage of realization by performing one or more of the experiments they have imagined. Instructors teach them to present this plan in a professionally coherent way that might convince investors. Communication skills, listening skills, willingness to rethink and compromise—instructors teach these things while guiding passionate dreams.

A few months into their idea translation experience, students are invited to give a public presentation. The presenters have practiced before classmates and selected class visitors. Now they have listeners with little incentive to attend other than to hear about the ideas themselves. Provoked by the risk of speaking outside the lab's tight-knit community, students polish their presentations, carry their ideas that much further ahead, and receive a grade or some sort of certificate.

Typically, more than half of students who start the program

wish to carry on with their project after their public presentation—and this is where the most transformative learning begins. The idea translation team works with these students to refine their plan, and the lab finds resources to send them somewhere else—to a different lab, a new city, a new country—during the summer to further develop their idea. The goal in all this is to encourage learning. It is a kind of bet on a student's capacity to matter in a dreamed-up way through imaginative thinking. If from this summer experience a company emerges or an art exhibition takes place or a nonprofit organization comes about, that's good news, though not the principal objective of the program. Students are expected to learn through idea translation regardless of whether they carry an idea as far as they initially hope.

Hugo Van Vuuren decided to take the "Idea Translation" course at Harvard in the fall of 2007. Around that time I had become involved in a group of artists, designers, and curators led by Susie Allen of ArtWise in London to brainstorm projects for lighting London during the 2012 Olympics. One fresh, unexplored idea was to use biochemical signals to make this happen. On hearing about this idea at the start of class, Hugo and a group of classmates expressed an interest.

Hugo and three other students studied the idea, looking into novel LED technologies and biochemical sensors. His group advanced well enough, it seemed, until midway through the semester, when the class prepared to brainstorm the "Lighting London" project. Hugo and his teammates, all with deep African experience—David Sengeh was from Sierra Leone, Zoe Sachs-Arrelano had spent time working in Namibia, and Stephen Lwendo was from Tanzania—announced that they'd changed their idea. They weren't interested in lighting London. They wanted to light Africa.

The team's new idea—no longer exactly an art idea at a frontier of science, but this didn't matter—put much more at stake than the one we had first asked them to consider. More important still, the idea was theirs. It aimed at helping millions solve an evening lighting problem that persists despite the fact that Africa has abundant energy resources and plenty of NGOs and governments standing ready to move existing technology, like solar or wind energy, to where it is needed. Where, then, is the irresolvable problem in Africa?

The problem is that energy infrastructure requires constant investment and local attention. It is one thing to own or have access to the relevant technology and expertise and investment resources. It is quite another to broadly and locally appreciate the value of preserving and developing it amid all the political, military, social, and economic challenges of sub-Saharan Africa. Recognizing that the lighting problem lay especially in the hearts and minds of African villagers was what made the idea of this group of undergraduate students so extraordinary.

Two of them, David and Stephen, had deeply relevant personal experiences with sub-Saharan Africa's lack of an electrical grid. While an adolescent, Stephen had stayed awake many nights after his family went to bed so that he could finish his homework by the light of a kerosene lamp, without depriving family members of the light they needed to cook, eat, and read. David's sister had, meanwhile, nearly lost her life because a doctor had been without electric lighting while helping deliver her baby at night. Kerosene lamps poison lungs and occasionally burn down homes in sub-Saharan Africa. Hugo and the others knew that if they could help find a solution to the problem of light at night in African villages, school performance would improve, crime rates would drop, and poverty would lessen for friends and families. Their idea mattered in a personal way that the idea of lighting London for the Olympics simply did not.

How would they move their idea forward? At the start, science guided them to an energy source, and art guided them to a novel approach to light emission. Hugo and his classmates discovered that relevant scientific research was happening right on the Harvard campus. Peter Girguis, then an assistant professor in the Department of Organismic and Evolutionary Biology, had figured out that microbial fuel cells, which derive their energy from the metabolic activity found in bacteria-rich soil, might be manufactured at very low cost and used to produce the small amounts of energy that basic lighting in developing countries required.

Peter's science was promising. The students felt it pointed to an approach that African villagers would more easily maintain and develop than other more obvious candidates, like solar and wind power. But finding energy was only half of the story. The student group needed also to find a low-energy-consuming, and safe, form of light, and in this they were encouraged by what they learned from British artist and entrepreneur Richard Kirk, a member with me of the "Lighting London" Olympics group. When Richard came to speak to the class, he told everyone of how artists were helping to explore the use of next-generation LEDs through light installations around the world. Might this technology be eventually helpful in Africa? The students proposed the idea, and it intrigued Richard, who remained a resource for the students for another couple of years.

So with these things in mind, Hugo and his friends set to work. Not long after he had announced his group's change of heart, Hugo and Aviva Presser, a talented teaching fellow in the class and a graduate student at the MIT and Harvard-affiliated Broad Institute, learned of a World Bank competition that invited proposals to light African villages—just as the students now intended to do. But the deadline for applications was imminent. With Aviva's help, Hugo and his classmates wrote up a proposal and submitted it on the last possible day. They explained in general terms how they would bring microbial fuel cell technology to African villages and help villagers begin to solve their lighting problem. A business model might even exist, they proposed, where they would manufacture products for the developed world, sell them at a profit, and use proceeds to drive infrastructure, training, and distribution in sub-Saharan Africa.

In the spring of 2008 the World Bank invited Hugo and Aviva to Ghana to compete with other finalists for the prize. To everyone's astonishment, they won—the only student-led group selected by the World Bank—and received a commitment of $200,000.

It was, of course, audacious for the students to have applied for the World Bank award with so little preparation, such limited knowledge, so few credentials. But it was completely appropriate to the artscience lab. In a sense the students were no less audacious than Fabrice Hyber, who believed he could share through contemporary art the experience of a stem cell creating a neuron. Or Ryoji Ikeda, who believed he could explore number theory through a soundless concert. Or James Nachtwey, who believed his photography could convey to those who live comfortable lives the pain of those whose lives are cut short by preventable disease. Hugo and his classmates had abandoned the original art idea we'd asked them to consider, and through their humanitarian idea created more convincing art.

Within weeks the students started a new nonprofit, Lebone, which means "light" in Swahili. And weeks after this, Hugo, Aviva, David, and Stephen took a field trip to Tanzania, supported by Harvard grants from the Heller Family Foundation, from Provost Steven Hyman, and from Dean Venky Narayanamurti of the School of Engineering and Applied Science. Finally, in August, they arrived in Paris for a workshop at Le Laboratoire. By this time they'd added a new Harvard undergraduate team member, Alexander Fabry.

After the summer of 2008, with the successes of Lebone, Le Whif, and MuseTrek, a buzz about the idea translation lab could be heard on campus. Hugo, having graduated, had signed on to work as director of the Harvard program. The intrepid Aviva was back as a teaching fellow. On the first day of class in the fall, a hundred students showed up. Previously, the number had been closer to twenty. Our sudden popularity presented us with problems of scale. Class time was just a small part of the teaching load. There was no limit on the hours a mentor might need to give, no way to frame the commitment it would take after the semester was over to fulfill the promise of each idea. And Le Laboratoire was starting its critical second year. My family lived in Paris. Could we do this?

After the first day of class we decided to exclude freshmen. Running back and forth between Paris and Boston, doing whatever writing and thinking I could manage on planes, I led an idea translation class that semester for forty students. This was still twice the previous number, and I guessed that the quality of class experience would go down. Hugo was one of the reasons this didn't happen. With Aviva, Hugo seemed to be everywhere that semester—on campus, on email, setting up electronic social networks, listening to students and directing their issues to me. His previous year's experience showed students what was possible, and they believed in the promise of the class from the start, probably more than any previous class ever had.

The sheer number of students helped, too. It turned out that a class of forty gave members more confidence to dream and take action than a class half that size. Learning like this was not a matter of plugging into some human source of encyclopedic knowledge, or even admiring an excellent example if you could not see concretely how it related to your own hopes. It was a matter of having an idea, of believing you really could translate this idea into a reality, and of receiving guidance and support as you fumbled along, came up with tests, abandoned intermediate hypotheses, and learned along the way.

Young innovators like Hugo and Aviva sometimes provide better guidance and support than do mentors with more seasoned skills, as critical as the occasional strategic advice of the latter may be. Young instructors express ideas in the raw, compelling way that students do when they are learning ferociously. The intensity of this give and take may have benefited Hugo and Aviva as much as it benefited the students. For a young creator, mentoring reinforces what you just learned, especially when you see it take form in the lives of other young people around you. First-time mentoring is a way of retracing steps made too rapidly to be fully noticed, and it builds confidence.

With the Lebone project moving ahead, Hugo became a kind of player-coach in the lab. Less than a year past graduation, he could show the new class what was possible if you found the resources to carry your idea forward. But Hugo was also learning the reality of keeping alive an ambitious, complex idea. He had helped start a couple of Internet companies while an undergraduate student and would soon be involved in the startup of others, but the challenges presented by Lebone were uniquely daunting. He had to invent a sustainable business model that delivered on a global humanitarian mission—something he and his teammates ultimately did by freely disseminating via the Internet their novel microbial fuel cell design. The audaciousness of the idea had already attracted attention in major national print media, including the *New York Times.* Harvard President Drew Faust frequently spoke of it in public addresses. The Development Office was holding up the idea as an example of what Harvard University students could do if you set them free to carry their dreams outside the classroom.

Obviously the idea translation lab benefited from all this resonance. By the winter of 2009, Hugo and his team were exhibiting the microbial fuel cell idea in a design exhibition at the Science Gallery in Dublin, bringing even more media attention to the startup. But unfortunately, the microbial fuel cells were not working as well as the team had hoped, and

Aviva, the lone senior scientist of the group, needed to figure out why. This would take another eighteen months and several unanticipated delays before Hugo, Aviva, and the others could finally travel back to Africa with functional, manufactured microbial cells and contribute to the first practical use of dirt-powered lighting in Africa.

As Lebone advanced, other projects began to flow through the idea funnel. One that emerged from the fall 2008 semester tried to capture the energy generated by soccer balls as kids kicked them around. This idea came about in a way that was unlike any I had yet seen in the class. Four young women formed this group. They spent most of the semester discussing, almost literally tussling, over their initial idea—something now forgettable—without consensus. Though all four were Americans, one had French roots, another Nigerian roots, a third Indian roots, and a fourth (Jessica Lin, later a leader in the idea translation lab) had Chinese roots. All were very articulate and held strong views. It seemed that their group would be the first to finish a semester without even figuring out what idea to pursue! We worried about it a little.

Then, two weeks before the semester ended, a new idea emerged that had no relation to the one they'd originally received. By the final presentation they were running from the back of the presentation hall, kicking soccer balls, which by the next summer would possess energy-generating magnets, and they would be telling us how they were going to build hope in the future of South African youth by giving them the chance to make power while playing team sports.

Several ideas relating to health in the developing world

emerged as well. One student group proposed new vertical farming methods for shantytowns, another pursued a novel approach to malaria prevention, another developed a form of viral radio specially adapted to the cellular phone capabilities of Africa. The Harvard Institute for Global Health started investing in the idea translation lab as a way to engage undergraduate students around ideas like these with clear potential benefit in the developing world.

Other design ideas that autumn were derived from biology, such as a project, which I patented with students over beer one night in Harvard Square, for transporting water in drought regions with an object that resembled a biological cell; or a novel performance-art project related to natural heart rhythms, drawing on the fascinating medical research of Ary Goldberger at Harvard Medical School. These projects elicited the interest of a second campus group, the Wyss Institute for Biologically Inspired Engineering led by Don Ingber, whose personal idea translation story is told in my book *Artscience.* From the institute's creation with a gift of $125 million by the medical entrepreneur Hansjorg Wyss in 2008, Don has championed the role of artscience in education and research and has backed this up with an investment in the idea translation lab.

Funding for the lab was not directed toward producing outcomes but rather toward providing the experiential learning environment that idea translation requires. The enthusiasm, the personal commitment outside class, and the obvious self-learning of students were what motivated investment by the Harvard Institute for Global Health and the Wyss Institute, much more so than demonstrable proof that these ideas would materialize as the students promised they would. Students learned to make the leap from what ignited their passion to what mattered to others in a complex, resource-constrained, technologically advanced world. This clear association of student passion with major global needs and opportunities attracted investors.

Learning how to survive the bewildering stage in a creative life when we realize that our most deeply felt commitments mean relatively little to others is a rare thing in a traditional education. Rare, because mostly we are never asked what our most deeply felt commitments are. The role of passion in what we do and why we do it gets forgotten. What do we seriously care about? Sometimes we don't even know. The typical choices we face—specialized as they become—are too dissociated from our inner passions to make us feel that wake-up-in-the-middle-of-the-night fervor. We are in a sense classified and processed by the educational system. As a literature student we may be asked to develop a theme in an essay, or as an engineering student we may be asked to propose a solution to a specific problem. Any of this can involve creativity. But it is a very different proposition to be asked what new thing relating to seventeenth-century English drama interests us so much that we would, if we could, convince someone to give us resources so that we could devote our lives to its pursuit. Or, even better, to be asked to come up with an idea, any kind of idea, which matters enough to us that we would devote our time and energy to showing others why it deeply matters to them as well.

If school didn't exist, an education would boil down to this question: What are we doing here? Learning how to answer this basic question in a way that motivates personal action is primitive indeed. However, in traditional schooling we give more attention to defining problems than identifying needs, even when it is in the spotting of needs that passion falls. Let young people figure out a need or opportunity, convince us it exists, and map out a path that will bring them support to pursue it, and they turn radiant. It's this radiance that drives investment in the idea translation lab.

My Harvard colleagues, notably Dean Narayanamurti, worried early on that this kind of educational project would not scale. Fulfilling the promise of such a class required too much out-of-class time. Yes, we were busy creating a larger artscience lab network to help curate ideas as they matured, but some student ideas, such as MuseTrek and Le Whif, fit better with the cultural incubation model of Le Laboratoire than others. At the beginning, Lebone fell into this latter category, almost purely an idea, too potential still to encounter in an exhibit and through this experiment with perceptions and beliefs. How would cultural incubation embrace Lebone? As it turned out, Michael John Gorman of the Science Gallery in Dublin asked Hugo and his classmates to exhibit an early prototype of their microbial fuel cell, which didn't work, and organized a conference around the idea, which did work and drew a huge crowd. When the students' idea went on stage in Michael John's artscience lab, an idea exhibit, it occurred to us that a more educational forum for exhibition was needed than the one in Paris. This eventually led to opening of The Laboratory at Harvard, a university exhibition space for student ideas in the arts and sciences.

The origins of The Lab go back to the creation of the Arts Task Force in October 2007 by Drew Faust just weeks after she became Harvard's twenty-eighth president. Led by Shakespeare scholar and professor Stephen Greenblatt, the Arts Task Force spent a year studying the role of the creative arts in a Harvard education. The group's final report argued eloquently for sweeping changes in the undergraduate curriculum, the development of new spaces for artistic expression, and the rehabilitation of existing campus facilities to that end.

The Laboratory at Harvard, which opened in November 2009, was a tangible result. It probably would not have come about as naturally as it did had Harvard's endowment not fallen with the economy in late 2008 and throughout 2009. The crash locked the administrative wheels of the university. Fretting over this unforeseen and dramatic swing in fortune, deans argued over resources, the administration laid off staff, and community leaders decried broken promises as university expansion projects were mothballed. The Laboratory at Harvard, which had small but growing resources and broad undergraduate support, stood out as a growth project at a time when nothing seemed to flourish in the traditionally centralized way.

A year and a half after graduating from Harvard, Hugo Van Vuuren became The Lab's managing director. His position capped an incredible ride of two years, beginning with his first idea translation class in September 2007. In the November 2009 opening of The Lab, the Japanese artist Ryoji

Ikeda performed a fascinating concert before an audience of seven hundred in Harvard's historic Sanders Theatre, while hundreds of faculty, students, alumni, artists, designers, and international visitors wandered around student exhibitions.

≡≡

By the time The Laboratory at Harvard opened, it had already taken us ten years to learn that what encouraged the dreaming of Harvard students could also encourage the dreaming of urban teens.

My wife, Aurélie, and I began the Cloud Foundation in 1999 using funds we had received from the sale of Advanced Inhalation Research, or AIR. This company had advanced an idea I had come up with at the beginning of my academic career: a particle specially adapted to deliver insulin through inhalation rather than by needle injection. The Cloud Foundation initially responded to the need for a high-quality artistic space in a central part of the city where kids' lives and artistic expression visibly counted. Cloud Place, located across the street from the Boston Public Library in Copley Square, became this kind of space. It encouraged the creative talent of teens by hosting artist workshops, public performances, and exhibition nights. For years Cloud Place had no other purpose than arts programming for Boston youth, in partnership with nonprofits around the city. By its ninth year the Cloud Foundation was reaching annually 10,000 urban teens, but the vast majority of them witnessed an isolated Cloud event and never came back. A much smaller group of teens, served by several nonprofits that sponsored theater, dance, art, and design lessons in rent-free space, grew up in and around Cloud Place, a safe place for artistic self-expression.

Each summer the foundation gave some of these kids the opportunity to travel to Paris, where we also sponsored art projects for teenagers. The Parisian teens made a reciprocal trip to Boston each fall. Young people in both cities had the chance to get to know one another, to see the power of artistic expression to bridge cultures, and to travel less as tourists than as citizens of the world. These cross-cultural exchanges proved to be among the most transformative opportunities the Cloud Foundation offered. During their time in Paris, teens from Mattapan or South Boston were perceived, and received, not as minority kids from a tough neighborhood but as Americans and young artists. Meeting teens who spoke a different language and yet, in so many other ways, appeared just like them proved refreshing and liberating.

By 2008, as Le Laboratoire was growing in Paris and the idea translation lab was gaining traction at Harvard, we saw an opportunity to tie everything together with a special teen program at Cloud Place. By then, other teen centers had arisen in the city, including the Boston Center for the Arts, not far from Cloud Place, and a beautiful downtown facility called Artists for Humanity. With other space options available in the city, we had to ask what real value we were adding. Bob Carson, a close friend and supporter of the mission and operations of artscience labs, suggested we encourage high school students in Boston to create in the same way that Harvard students were creating—to pursue art and design ideas that produced cultural, humanitarian, educational, even commercial change.

Cloud Place could produce opportunities for inner city teens that were unavailable elsewhere in Boston, and our uniqueness would attract resources to expand the program. Bob argued that the foundation's new mission should be to create an idea translation lab at Cloud Place to seed the kind of passion for idea exploration that is so hard to embed in the public school system. Each year the program would reward in some way several ideas emerging from the minds of kids in Boston.

In the summer of 2008 we recruited Carrie Fitzsimmons from the Institute of Contemporary Art in Boston, where she had been chief-of-staff to the director, Jill Medvedow. By October, thanks to the advocacy of real estate developer Mark Maloney of Boston World Partnerships, we had the support of Boston Mayor Thomas Menino. By January, Mayor Menino was announcing a new ArtScience Prize in his State of the City Address—an annual grant award to one or more groups of Boston public high school students for the best art or design ideas to emerge at a frontier of science. By the following March, Carrie and her team at Cloud Place had studied our program at Harvard University and were starting to tailor a curriculum for high school students. We recruited a jury of thirty artists, scientists, and entrepreneurs and met with the majority of them over a dinner at Cloud Place to brainstorm the first set of ideas that we would propose to the kids the following fall.

That first year we chose the theme of neuroinformatics, a frontier of science that seemed easier to grasp than many other frontiers we might have targeted, like the biochemistry of interfering RNA. Ideas that emerged from the jury included the practical "argument resolution hat," articulated like this:

"You put on a cap, called a 'brain computer interface' (BCI), and your friend puts on one, too. You're in a bad argument and there seems no resolution. Well, good news. The hats are going to figure it out. Perhaps a series of potential resolutions are shown to you both, twenty different ways to compromise, and when the two hats show the greatest agreement, you've found your resolution. Wouldn't that be a great technology? Think of all the money it would save in tough negotiations!" But there were also more abstract ideas, such as the "art of the forgotten," which went like this: "We forget more than we remember. What this means is that we learn all our lives and we end our lives remembering just a few things. Only a few things really stand out. We tend to represent these things in speech and, if we're artists, in our art. But what if you turned it around? What if you could make art based on everything you forget? What would that mean? For dance, music, visual art?" Other ideas were simply fun, like "dream player": "What if you could wear a cap at night, and the electrical (EEG) data gathered by the cap somehow recaptured your dreams, and in the morning, while you're brushing your teeth, you could watch your dreams, a kind of fascinating morning television! Would you tune in?"

On the more general matter of "tuning in," we had our doubts. We hoped high school kids would embrace these ideas with passion, but would they? The city's annual school dropout rate had been around 10 percent since the early 1980s—nearly 50 percent over four years in many schools. Located in one of America's richest centers of higher education, the Boston public schools seemed cut off even from local education and professional opportunities, never mind national or international resources. Would Boston teens find any of the Art-Science Prize ideas of interest? Would they pursue a dream as seriously as highly selected university students did?

In the spring of 2008 we started to share a few of the ideas with students at Boston high schools. Meanwhile, Carrie and the staff worked at building the art workshops and brainstorming sessions of the idea translation lab at Cloud Place. This program would require teens to show up one or two afternoons a week from October through mid-January, make proposals, and present them in various rounds of evaluation. At the final presentation, with the mayor participating, the Art-Science Prize would be distributed among the teams with the best ideas. The top group would travel to Paris for the summer Innovation Workshop at Le Laboratoire, where they would develop their idea further with students from Harvard University and other schools in Paris, Brussels, Dublin, and other cities. Not every Boston group would win a prize, of course. But the Cloud staff made a commitment to work through spring with all groups that remained passionate about their idea, whether it was funded or not.

Carrie and her team also reached out to industry, thinking that as students advanced they might be good candidates for summer jobs. She pointed out to high tech companies, especially those without much of a civic tradition, that they could begin serving city teens by employing them over the summer. The biotech company Amgen was first to step up to support the prize.

Carrie and her team guessed that Cloud Place's idea translation lab could handle 130 teens, perhaps a few more. But what evidence did they have that the program would attract urban teen interest? Very little. In the fall of 2008 we'd asked several kids involved in a teen curator program at Cloud Place if they would be interested in working with us on an idea related to MuseTrek, and four of them raised their hands. Hannah Cummins was in her junior year at Brookline High, and Jake Giberson was in his senior year. Jeffrey Cott was in his senior year at the O'Bryant High School, and Isis Cortes was in her junior year at Fenway High. Two of the Harvard student founders of MuseTrek, Tarik Umar and Roee Gilron, agreed to come regularly to Cloud Place to teach the kids the basic principles of the project.

The MuseTrek idea, which I describe more fully in Chapter 5, had emerged a year before out of the idea translation lab at Harvard, under the leadership of Mishy Harman and several other undergraduates. In the fall of 2008 MuseTrek was still an early-stage idea about creating and sharing adventures, or treks, that could be downloaded on a smart phone by others as they visited a cultural venue. A year later, in the fall of 2009, the idea would become a practical technology and the basis of a startup company.

Our pilot teen group made some treks, and in early 2009 they came up with an idea they called CityTrek: a tour guide for Boston neighborhoods. In the summer of 2009, just before our Paris workshop, they tested it out with twenty or so friends in Boston's Back Bay. It went well enough, but you couldn't exactly say that the teens were passionately developing a personal idea. Then, in August 2009 Jake, Hannah, and Jeffrey came to Paris for the summer workshop. They joined fifteen Harvard students, eleven design students from Paris's

Strate College, and six students from ERG (Ecole de Recherche Graphique) in Brussels.

The workshop experience involved a few of hours of presentation and discussion each morning, with afternoons free to develop ideas, work in groups, and interact with other teams, knowing that by the end of the workshop they would all need to give presentations, open to the public, which would convince us that they could carry their ideas even further. Add to this relatively free-form workshop structure the complex mix of students from various countries and the competitive atmosphere of idea development, and you might easily conclude that this workshop would kill the motivation of our Boston high school students, and with it the promise of the Art-Science Prize.

A day into the workshop I met for a long afternoon session with the Boston teenagers. They asked to speak to me alone. A few seconds of silence. Go ahead, I said. What's on your mind? They felt pushed, they said, not free to make their idea whatever they wished it to be. Perhaps it was the arrival at Le Laboratoire, or being in a new city, or maybe it was just the vision of other students, so excited, motivated, sure of themselves. Whatever it was, these Boston high school students had a message for me. We'd chained them to MuseTrek, an idea developed at Harvard University by others. And while CityTrek bore their own local stamp, it was pretty indistinguishable from its predecessor. So what really was their idea, anyway?

We discussed their plans for the future. We talked about what it meant to pursue an idea, the risks and rewards this involved, the constant need to convince others to care enough to help you out. They smiled. This was precisely how they wanted to build their futures. By the end of the day Jake and Jeffrey (Hannah continued to work with the original Muse-Trek vision) were off and running with a new idea. They were going to create an Internet and handheld application to orient students at the Massachusetts College of Art, where the two were planning to go to school in the fall. Jake and Jeffrey had come up with a riff on the MuseTrek score that addressed an opportunity they alone saw.

For a week their idea kept Jake and Jeffrey from getting much sleep. They spent hours each day with Hugo, who had assumed responsibility for managing the workshop—getting the students to show up each morning, planning afternoon games, evening dinners, following ideas and helping them take shape. By the end of the week Jeffrey and Jake were giving the funniest and possibly most convincing presentation of the workshop. The designers from Strate College said they couldn't believe what they saw. High school kids? They were fabulous.

Back in Boston, Carrie began to recruit mentors to oversee the curriculum—someone had to teach, after all. Her recruitment materials advertised the program and its goal of helping high school students dream in ways that mattered to as many people as possible. She and her team shared the ideas that the jury had put together, and they used the example of Jake and Jeffrey as evidence the program would actually work.

Then a second extraordinary thing happened. Carrie began to receive CVs from remarkable teaching candidates. One was finishing her PhD in brain and cognitive science at MIT, two were successful architects, one was to have an exhibition of his work at the Tate Modern later in the year, another had a company that made works of art out of molecules, to help people understand molecular function. All were practicing artists, all were interested in science, and most had some teaching experience. I met with this group a couple of times and invited them to my class at Harvard in the fall.

Carrie and her new ArtScience Prize team invited all interested high school students to come to the Boston Ballet Center after school on September 21, 2009, to learn about the prize and select an idea they might develop with us in the fall. The week before the event, the team went around to about a third of the Boston public high schools, speaking to around 4,000 students. In an encouraging sign that excitement was growing, school leaders began calling Carrie at Cloud Place, asking for the team to come to their school.

But would the kids themselves volunteer to compete for the ArtScience Prize? Carrie and her team prepared the Boston Ballet's main practice ballroom. The twenty-four art and design ideas in neuroinformatics that the jury had developed were displayed on posters, creatively illustrated by artist-mentors and arrayed around the room with white paper that we'd stretched across the walls. I ran to class at Harvard that morning, then came back to the ballroom in mid-afternoon, thirty minutes before the kickoff. People had already started to arrive. They were sitting on the floor, mulling around the posters, taking notes. By the time we started addressing the group, nearly three hundred Boston kids filled the room.

After a twenty-minute introduction that clearly stated the tremendous work this program would involve, we asked the teens to stand up. Whoever didn't wish to pursue the program should leave, as next we would be going over the ideas and getting to work. Everyone stood up. Only a few left. Most were talking excitedly—a good sign. As I reviewed the ideas, there was energy and laughter, and students occasionally talked back to me. By the time it was over, with attention riveted all around, the students got back up on their feet and wrote their names under the projects that interested them, signing with a flourish. As we neared the end, one yelled out, "I'm excited!" There was some whistling—as if a ballgame were about to begin.

Two weeks later about 200 students showed up at Cloud Place to enter the afterschool idea translation lab. Approximately 100 students were accepted into the program and continued several weeks with the program, learning art and design techniques and twenty-first-century skills of communication, group collaboration, and brainstorming. At the four-month mark, around sixty students had succeeded in developing and presenting fifteen different breakthrough art and design ideas in the field of neuroinformatics.

Much work remained ahead to identify the winners of the prize and to work with each group of students in a way that would direct their passion toward an increased capacity for self-learning. But Jeffrey and Jake's enthusiasm wasn't an outlier. Urban teens were every bit as willing to dream and to learn through idea translation as were accomplished university students. And in the months that followed, other teens and university students began learning through idea translation labs in Oklahoma, Dhahran, and Paris, inspired by the universal promise of learning through dream realization across boundaries of culture and knowledge.

≡≡

If the idea translation lab has served students well, those who finish it are still passionate about remarkable ideas that have emerged—new ways of lighting Africa, of eating, of understanding culture. To fail to help these ideas come to fruition, to let them fall by the wayside because they aim too high, too soon, too youthfully, is to dissuade students from the promise of the future, and to betray the promise inherent in this kind of education. Indeed, the rest of this book is about an educational commitment, a kind of idea translation endgame, to embrace ideas so daring that they remain without any clear cultural, commercial, or humanitarian value and to help them attain external value through cultural incubation.

⇒4⇐

CULTURE

In the cultural exhibition programs of the artscience labs in Dublin, London, and Paris, art is inseparable from the creative process. Works-in-progress exhibited in these and other contemporary art and design galleries may originate in educational projects and, later on, lead to commercial or humanitarian innovations. But the core value of cultural exhibition resides in the dialog it inspires around the creative process itself, whatever its origin and wherever it might lead. This proves equally attractive to experimental artists, such as those described in this chapter, and to visitors seeking more intimate and participatory cultural exchanges than are possible in traditional museum exhibitions.

At the Wellcome Collection in London visitors are asked to make sense of what science, and particularly medical science, means to their daily lives. Months before opening the gallery, Director Ken Arnold held parties and conversations around the city designed to appeal to the "incurably curious." This format of participation now informs all programming at the Wellcome Collection. Lunchtime, evening, and tour events encourage visitors to participate in debate ("Should we stop breeding?" was one recent provocative theme), or simply in the mystery of fleeting moments (exhibitions, such as a popular one called "Skin," or another entitled "Darwin's Inheritance," went up and came down in a single evening). When you come to the Wellcome Collection, you are not handed universal answers but, through the fusion of art with medical science

—along with anthropology, evolution, psychology, history, sociology, and much more—you participate in the pursuit of answers.

Michael John Gorman takes a different but equally effective approach in Dublin. His Science Gallery invites the public into the process of science exploration, frequently through art and design. In his artscience lab the public explores infection by gowning up and participating in a form of theater, or they think about the problems of crowd control by playing computer games that build virtual crowds.

Valuing art through participation in the creative process is not exactly mainstream. We know better how to understand and value art as outcome. A September 2008 Sotheby's sale of artist Damien Hirst's works netted over one hundred and forty million euros. It is hard to argue that the money invested in creating these works, even with an exceptionally high monetary value assigned to Hirst's time and creative input, is anywhere near what the art markets were willing to pay for what came out. That Hirst's creative process was influenced by a fascination with medical, pharmaceutical, and forensic science was not an obvious factor in setting the price of what he made.

Can the value of art as process ever really match the value of art as outcome? Next to outcome, the creative process can seem like a footnote. But to say that what comes of the creative process, whether it is a pencil drawing by a five-year-old or a jeweled skull by Damien Hirst, is more investment-worthy than the process that led to the created work is, among other things, to ignore the fact that culture changes. Valuable as works of culture are today, the works of tomorrow will be more resonant with the opportunities and worries of the future. Outcomes of art remain affixed to a time and place, however usefully or meaningfully, while the artistic process does not.

Labs, as environments that support, frame, and rationalize the value of creative process, are a fragile link to the future, and therefore critical to the future relevance of cultural expression. Assuming they operate effectively as catalysts for discovery, they should remain as mysterious, unpredictable, and promising as curious children—a vision I explored, under the pseudonym Séguier, with two experimental graphic novels during the first two years of Le Laboratoire: *Niche,* written collaboratively with novelist Jay Cantor and illustrated by New York artist Daniel Faust, and *Whiff,* illustrated by the Manga artist Junko Murata. *Niche* and *Whiff,* written with the intention of understanding and experimenting while also sharing and inviting, became "process" catalogs to accompany "product" exhibitions.

≡≡

Two days before opening Le Laboratoire, Fabrice Hyber put the last touches on his *MIT Man,* a body composed of fruit and vegetables conceived to map the biological universe of a stem cell. Beneath the catwalk on which I stood was Fabrice's giant inflatable hourglass, which you could jump into with a short leap. If everything went right, you would fall through a sleeve of stretchable polymer onto a conventional bed mattress.

Fabrice glanced my way and asked if he might paint the floor. It wouldn't cost too much, he said, wouldn't pose much of a cleanup problem. The paint would wash right off. The cement was gray and looked just fine beneath the white exhibition walls he'd arranged in the form of an arrowhead, dividing up the space into three sections, two outside the walls, one inside. He wanted these three sections of floor painted red, blue, and green. "I've done this kind of thing before," he was saying, incredibly. "We need to do it now or it will never dry."

Seeing how delighted Fabrice was by the idea, we simply did it. Not that we believed him when he said it was going to be an easy operation, like rearranging the lighting. There just did not seem to be much of a choice. Something undeniable hung in the air that afternoon. Fabrice's idea was alive and vulnerable around us and we couldn't resist it or ask what it meant beyond the plain facts of it—red, blue, green.

By evening Fabrice was flopping his mop around in a big bucket of red paint, laying the oily liquid on thick. François Maurice, who oversaw the exhibition's construction, walked onto the catwalk and stared blankly at the colorful floor and at Fabrice with his playful mop. This was going to cost a fortune, he said. The place smelled like toluene for a week.

A month later, around thirty investment bankers came to Le Laboratoire. The BNP Paribas had asked the lab team to organize a workshop on innovation. Of course, innovation in an artscience lab and innovation in a bank were two different things. Innovation at BNP Paribas had a quarterly cycle to it. You measured it quantitatively, exactly. To innovate was to surpass financial goals in an industry where many people with competing interests fought against you. Ambiguity led to failure. Precision helped you meet targets and be rewarded. What we did at Le Laboratoire seemed closer to play. But for leaders of BNP Paribas, an ability to see opportunity outside traditional constraints in difficult economic times held enough potential value to warrant a full day's gathering in Paris.

During the workshop there was talk about idea translation, about obstacles to innovation at Le Laboratoire and at the bank. Then the bankers split up in groups and did a creative design project. They made roller coasters with insulation tubing, tape, empty yogurt containers, and marbles. With music playing in the background, they crawled around Fabrice's paintings and installations trying to come up with the funniest, fastest, surest designs.

At one point a banker stood up next to one of Fabrice's paintings that had an image of a brain on it, with some sketches of wine bottles, bubbles, and an apple. "This isn't innovation," he said. "Anybody could have done this. My little boy, for instance." Having passed an hour on his knees making a fancy roller coaster design, a kid again, this successful businessman had perhaps made more of a conceptual leap than he realized.

To stop seeing innovation as what's done and to see it as "doing" is perhaps the secret to understanding innovation at all. Not that we knew well at the time how to involve our visitors in what we experienced as innovation—a passionate, experimental meeting of art as process and science as process—even with our roller coasters.

≡≡

Ryoji Ikeda is rigorous in each artistic idea he pursues. You have the sense while talking to him that his mind harbors some inexhaustible idea mulled over for a very long time. An idea that leaves him politely courteous when the conversation strays too far away from his artistic work or from topics of pure math or from notions of the sublime; a very ancient idea, akin to self-sacrifice, that produces in him a patient weariness, a silence that contrasts amusingly with the youthful bounce in his step, with his way of making fun of what he has just said, or is about to say, as if he hasn't thought about it that well after all.

Ryoji says he did not enjoy school. After high school in Japan he worked in sound and music performance. He also wondered about the relationship between music and mathematics and began to read deeply on the subject. The idea Ryoji developed with us at Le Laboratoire grew out of this.

When we met for the first time at the Café de l'Epoque, Ryoji said he was investigating number theory, and he shared some images of a concert he'd just performed at the Centre Pompidou. He called it *datamatics*. The Festival d'Automne was interested in co-sponsoring an experiment with Le Laboratoire, and Caroline Naphegyi wanted to enlist Ryoji. I asked him if he had yet worked with a leading number theorist. He hadn't, but if we might help him meet an elderly Japanese mathematician he had in mind, he would be delighted to see what came of it.

A couple of months went by. The Japanese mathematician turned out to be retired. So I asked around Harvard about number theorists on campus, and many people mentioned the name Benedict Gross. I'd spoken to Dick a couple times when he served as dean of Harvard College, so I sent him a note.

Would he be interested in meeting a Japanese digital composer? I sent a URL as background. Yes, very interested, he wrote back, without delay. It turned out that Dick had a long-standing passion for music, had even thought as a young man about pursuing a music career, and continued to play regularly in a string quartet. We set up the meeting and Ryoji came to Cambridge in January 2008.

To write that Ryoji and Dick immediately took to each other, or that they discovered over the next months a common language intermediate between art and mathematics, is to focus instantly on a moving outcome of the experiment and to lose sight of the process by which it occurred. It seemed that Ryoji and Dick each grew almost younger through this experiment. A kind of symmetric transformation began the first evening they met, Ryoji flying in from Paris, me picking him up, driving him to my office, Dick arriving, his fantastic, immediately deferential humor making everyone relax. Ryoji pulled us into a little conference room next to my office, turned on his computer and started his *datamatics* concert, images projected onto a screen, lights out. The experiment started this way.

Later on that evening, as we walked through Harvard Yard on the way to dinner, Dick observed that all number theorists were finally Platonists. This off-hand remark tickled Ryoji. He asked Dick to repeat what he'd just said, as if he'd heard what he'd long assumed to be true and couldn't get over it.

The emails that went back and forth between Ryoji and Dick over subsequent months created a prose mirror of the experiment, which Ryoji arranged in the form of a beautiful catalog of the exhibition, V≠L. Ryoji's letters began innocuously enough, a few days after he left Cambridge, in January 2008:

> Dear Benedict,
> It was truly great to see you in Boston, thank you very much for spending your precious time. All you said is still resonating and expanding in my head . . .

Dick wrote back the very next day to say:

> I can still see many of your images in my head. I love the datamatics project that you left me, especially the notion of a vast data ocean . . .

Ryoji mailed a collection of CDs. Dick listened to them and wrote back another email, this one on February 10, 2008:

> Your work is beautiful. I wish I could play some of the compositions for string groups, but my quartet is probably not up to the technical demands . . . As for me, I am busy at work. But I look forward to seeing you in the fall.

Ryoji thanked Dick for his "kind message" and made clear his plan for the winter and spring, how he would like to collaborate, artist with scientist, on the exhibition for the fall:

> I have been preparing some questions mostly about your own standpoint on aesthetics and philosophy of mathematics. I understand you must be busy, but if we could make some email exchange on that issue, it would be really appreciated. I would like to put our correspondence as artistic process of this project in the catalogue, which must be a crystal-clear art

object containing a sublime subject between art and mathematics.

For a while, I am also busy preparing my solo exhibition in Japan until the end of March. So how is your availability during April-May? I cannot say anything about how our conversation will be—maybe boring for you (in this case please tell me) or maybe very productive. Anyway I am very excited to keep it open/unpredictable . . .

A couple of months later, on April 22, Ryoji wrote back to Dick, now giving his "first statement of the project to the press":

As an artist/composer, my intention is always polarized by concepts of "the beautiful and the sublime." To me, beauty is crystal; rationality, precision, simplicity, elegance, delicacy.

The sublime is infinity; infinitesimal, immensity, indescribable, ineffable. The purest beauty is the world of mathematics. Its perfect assemblage amongst numbers, magnitudes and forms persists despite us. The aesthetic experience of the sublime in mathematics is awe-inspiring. It is similar to the experience we have when we confront the vast magnitude of the universe, which always leaves us open-mouthed. The aim of this project is to engage in dialogue with mathematicians to find a common language on aesthetics.

Ryoji added:

This text is simple but quite personal. Whether it's correct or not, this is my standpoint for this project. So, before I speak about many things, I wanted to give you this text first. I hope that we can start from this. I look forward to hearing from you.

The next day, Dick wrote back:

> Good to hear from you. I have been busy with work too. I like your initial statement. There is really no right or wrong here. As you say, you are in the process of contemplating the sublime.
>
> I think that most mathematicians believe that the truth in mathematics is something that we are honored to discover. We don't create it, we simply uncover pieces of it. And once we see it, it is always more beautiful than anything we could have imagined.
>
> I'm attaching a short piece I wrote for a book that is coming out next year, entitled "Faces of Mathematicians." I don't think it is particularly relevant to your project, but it will give you some idea of how I work. In any case, I look forward to our conversation, and to seeing you in person in Paris at the end of September. It will be a pleasure to work with you.

Ryoji wrote a new email the next day. Here he asked the questions he'd promised back in February, wanting Dick's thoughts on Platonism, infinity, formalism, intuitionism, transfinite numbers, sheaf theory, the Riemann Hypothesis, and many other things.

> Sorry, it may be a bit stupid to ask you like this, but if too technical, at least I want to grasp the whole (sorry, my mathematical ability is very poor) . . As I stated, "the beautiful and the sublime" is an important part of aesthetics. After Burke and Kant, this issue is somewhat too trivial for artists, but I still see a great potential to drive me, I want to hear your thoughts. Would you show me one of the best examples of

the beautiful and the sublime in mathematics? For example, to me, the beautiful is $e^{i\pi} + 1 = 0$, the sublime is the continuum hypothesis. Here, I put some keywords, which have been attracting me for years. These are the seeds to develop our conversation, could you respond? I think all's quite heavy and profound. So I just want to start it casually (with my full respect).

Dick reflected on this for a few days then wrote back thoughtfully about his own work, and about some of the issues raised by Ryoji:

I have never met a great mathematician who was not a practicing Platonist. The subject is so beautiful; it is far beyond the powers of human creation. Of course, it would be nice to step outside the cave once in a while (perhaps artists can help us do so). I think the arguments about constructive mathematics, intuitionism, etc., are largely sterile. Working mathematicians all know that the subject holds together, even if the analytic philosophers cannot formalize it. If someone shows that some statement we are actually interested in is unprovable, we will address it—much as our predecessors defined the notion of convergence in the beginning of the 19th century, when issues arose with the use of divergent series.

As you know, the notion of the size of infinity and subsets of the real numbers are closely connected. Your questions "How many points are there in a line?" and "What is the number of numbers?" get to the heart of this. I hope you will be able to convey the wonders of Cantor's discoveries in your installation.

Remember, at his time the general belief was that any un-countable set of real numbers would contain an open inter-val! Cantor showed that his set was totally disconnected—be-tween any two points there is an open interval not in the set—and was still uncountable. So it's not a question of vol-ume in the traditional sense. There is a lot of room down in the small world of dyadic or triadic expansions that you might help us see (or contemplate). I haven't been able to ad-dress all of your questions, but I hope this helps.

The experiment went forward like this. A record of the creative process emerged. It showed up with these and other words on paper. But not just on paper. This time the process came to life in the exhibit too. Ryoji described in a June email what it would look like:

Two large photo prints . . . in parallel, 30–40 cm (1–1.3 ft) above the floor as if they were floating. Their proportion is unusual, 1 m wide x 7 m long (3.3 ft x 23 ft). One is filled by 10166688 digits, which is a number (that consists of 10166688 digits). I began test prints today at le lab, trying from the number e. However, I am interested in using a totally random number, which is not mathematically reducible. The reason is simple: I want to see it. Also, according to Cantor, most num-bers are transcendental. The majority of all numbers are numbers which have no pattern. Perhaps those numbers are mathematically not interesting. But at least, the existence of any kind of number is a mathematical truth, moreover I am very curious to bring such a mathematically insignificant number to our world from the mathematical world. Otherwise, we human beings have no chance to ever encounter this particular number! Also, I can say that the number might be the first (or middle or wherever) 10166688 digits part of an infinitely random number. The number is a singular point of a real line, no different in importance from zero or one in a democratic sense . . . But how can I find such a single pure (not pseudo-) random number which has 10166688 digits?

Ryoji figured it out a few days later, as he wrote in another email to Dick:

Finally I decided to use the single number which consists of 7261920 digits, randomly generated by computer (I gave up finding the way to generate the number TRULY random, I think it's impossible). The result of the photo print was something I've never seen in my life. Today, I finally solved the major technical problem, and I am now preparing to construct and brush up physical materials. Since the schedule is getting tight, I was asked to decide the title of the exhibition. As was my initial idea, I'd like to title it $V \neq L$. I believe it's true. And I like it because it is the simplest expression about something very important about mathematics itself. How do you think?

One of the earliest critiques of the exhibition came from *The Guardian* a week after the opening in October 2008. The *Guardian* critic wrote, "From anything but extremely close up, the panels are fuzzy concrete-gray prints. Close up the viewer becomes mesmerized by the mass of 0.8mm-high digits, which are daunting in their vastness and precision, but hint at a common aesthetic language shared by art and maths." Ryoji Ikeda's experiment was among the most impenetrable projects of our first years. "The exhibition . . . is simultaneously baffling and fascinating," according to a December 9, 2008, critique in *Artslant*. Ryoji wanted a "soundless" concert at Le Laboratoire, an idea not lost on the critic of the Asian Art Press *World Sculpture News*. "Decibels take life and run up against each other; sounds are oppressive or liberating; is it resonance or silence that we're hearing?" "Surrounded by silence," wrote the *Artslant* critic, "the use of light here mimics the inexpressible perfection of numbers in sequence, a perfection that words cannot express."

The exhibition seemed to end local speculation about the purpose of the lab. Ryoji had found an artistic language to show what we were about. But Dick Gross may have had the last word. In an email he sent just before the opening of the exhibition, he wrote to Ryoji:

> When I was in Spain this summer we visited the Neolithic caves in Altamira, which have wonderful ceiling paintings of bison, horses, hunting scenes, dating from about 20,000 years ago. Picasso said, "After Altamira, all is decadence." One could say the same about mathematics, after Archimedes.

There was a little sadness in this, as if Dick's anticipation of the exhibition that lay ahead couldn't match the pleasure of the creative process that lay behind. But then a year later I happened to cross paths with Dick on the Harvard campus. The *datamatics* concert had come to the university for the opening of The Laboratory at Harvard and had just left. Surprisingly, Sanders Hall had been filled to capacity, and we were all still talking about Ryoji's success. I could see Dick was in a hurry.

But his brief words to me said it all: "Am exchanging emails again with Ryoji! We're onto something new!"

≡≡

Shilpa Gupta could not have been more unlike Ryoji in her approach to Le Laboratoire. Science was not her particular interest. That her art might lead to an exploration of some frontier of science, she could imagine. But she wasn't sure. We had to work hard to enlist her. This was now our second conversation, and each time Caroline Naphegyi sat silently, supportively, at Shilpa's side. Shilpa trusted Caroline's artistic judgment, having worked with her in the northern French town of Lille—and probably wouldn't have been meeting with us for a second time were it not for Caroline.

Shilpa said "fear" interested her as a subject of artistic inquiry—notably the fear that others imposed on you, formed in you like a bomb, and used to break you down. She called it terror, a word she sometimes used with precision, as something that could be wielded. I understood that her experiment would be complex, unpredictable, rich, and without end.

Shilpa had grown up in a large traditional family in Bombay and lived through the democratization of her country. She had watched as what seemed like a national spirit of altruism gave way to sectarian violence that split people into antagonistic groups, Hindu and Muslim, reactionary and tolerant, those who looked back and those who looked ahead. You might start out in one group and end up in another, or you might be strong in your beliefs. Shilpa suspected the distinction related to how you handled the viral, invasive, omnipresent role of the media and the technologies it used to influence our brains. Might the explanation be found in the biological, in the genetic, in something that predisposed us to terrorizing others, or to succumbing to it?

Our first conversations brought us to the field of neuroscience. I shared names of scientists with whom Shilpa might discuss her question. She began with a Spanish scientist at the University of Santiago who studied how and why the brains of monkeys process visual information differently. The subject had interested Shilpa, but after speaking with the scientist she decided the work was too far removed from her artistic question to be of interest. We arranged a second meeting in early 2008, this time with Dean Mobbs of Cambridge University, who uses magnetic resonance imagery to study the cognition of fear. Shilpa and Dean met in his Cambridge office, and while the conversation approached more closely the subject of terror, it ended without an obvious path forward.

Shilpa fell silent—hard to reach by phone or email. We began to wonder if the experiment would happen at all. Then, in the spring of 2008, Shilpa agreed to travel to Boston from her home base in Mumbai in order to meet other scientists. I introduced her to Pamela Sklar of the Broad Institute, who studies genetic diseases of the brain. Pamela explained that our genes probably influenced any physiological condition you could think of but only a few human diseases seemed to have a clear link to a small and definable set of genes. We talked about the genetics of schizophrenia and Alzheimer's disease. Then we opened up the conversation to more speculative subjects. Might there be a gene that made you a Republican or Democrat? Pamela laughed at Shilpa's question. Yes, probably you could find something there, though you'd never get an ethics committee to approve a study to look into it, and it wouldn't make good science anyway. Scientists do best when there is not a vested interest in the results of their experiments, and political opinions in the country around that time, perhaps at any time, were too deeply felt. What about fear? Pamela said there were scientific studies that pointed to a genetic basis for fear, and Shilpa might study the genetics of fear in the course of her art experiment, if she liked. Shilpa was very interested in this. We might follow published protocols of experiments on fear and design them to answer Shilpa's questions.

There are a few tests by which scientists today study the genetics of fear. One involves measuring physiological reactions—heart rate, sweat rate, facial movements—while human subjects watch frightening images, such as those they might see in a horror movie. Physiological reactions are compared with genetic similarities to determine whether a statistically significant genetic correlate exists. If fear is genetic, identical twins, who have the same genetic makeup, should react to the frightening images in more similar ways, while fraternal twins, who share half of their genetic makeup, should react less similarly, and people without any genetic relationship should react least similarly to the frightening images.

Carrie Fitzsimmons had been with Cloud Place for a little over a month when the idea of the genetic experiment came up. Why not involve young Boston public school students? It might help us understand challenges and perhaps fruitful ways in which ideas would move back and forth between the emerging idea translation lab at Cloud Place and Le Laboratoire. Carrie began to explore how Shilpa's genetic experiment might provide creative opportunities for the urban teens with whom we worked. Shilpa had previously visited Cloud Place and appreciated the idea of engaging urban kids in her work.

Carrie came across the research of the psychologist and neuroscientist Mahzarin Banaji, who teaches at Harvard. Her work, popularized by Malcolm Gladwell's *Blink*, had a noninvasive way to get at some of these same questions of fear. It was called the Implicit Association Test, or IAT. Carrie brought the IAT to Cloud Place and started doing experiments with identical and fraternal twins. When Shilpa came back to Boston in late November 2008, we introduced her to Mahzarin. And now, only two months before her scheduled exhibition, Shilpa had found her scientist collaborator.

Shilpa came prepared to her first meeting with Mahzarin. Soon after we entered Mahzarin's office she took out her tape recorder. We had a second conversation at Mahzarin's home a couple days later, and Shilpa did the same thing. We discovered her to be a persistent researcher.

Yes, the IAT could be used to assess fear, Mahzarin explained. "In brief, the test measures how fast and how accurately we respond when putting two things together—for example PARIS + GOOD and PARIS + BAD. If I am faster at PARIS + GOOD than PARIS + BAD, then my 'implicit' or 'automatic' liking for Paris is positive. If I am faster at PARIS + BAD than PARIS + GOOD, then my automatic feeling for Paris is negative." Mahzarin said that very negative feelings toward someone or something, as assessed by the IAT, tend to correlate with a fear of this someone or something, hence the utility of our experiment.

This was all very useful, but it didn't directly answer the questions that worked on Shilpa's mind, the kinds of questions the other scientists we'd met had found difficult to answer. Their dialog was again recorded in the catalog for the exhibition:

Shilpa: Can you say something about fear being biological? Is it something you're born with?

Mahzarin: Of course, the mechanism by which we learn fear is in us. We evolved to be able to experience fear, because fear is a very useful emotion. It helps us to know what not to approach, not eat, not mate with. And it may be there are certain things we fear that we don't even need to learn about. Monkeys who have been raised in captivity and have never seen a snake, shiver when they are first confronted with one, suggesting that such fear is not learned. However, the vast majority of things we love and hate, fear and approach, are learned because our culture—families, friends, media—give us those experiences. If I grow up on the West Bank, Hamas is my defender; if I grow up in Israel, Hamas is my enemy. Who fears what, in that sense, is fully learned and acquired through experience.

Shilpa: Where does terror come from?

Mahzarin: Terror comes from a threat to our well-being. It can be to our physical well-being. The recent events in Mumbai, 9/11, and other acts of terror that we associate with terrorists are the most obvious, direct examples of terror that comes from harm to one's physical self. But as a psychologist I'm also interested in terror that is produced because the harm is expected to be to my psychological self, my sense of worth, my dignity as a person. And the hierarchies that mark all societies give us a way to study this. Terror may be a strong term, best reserved for particular acts of violence, but something like it may be experienced even when our bodies are not at risk but our way of thinking is threatened. When one's religious beliefs or moral beliefs are threatened, something akin to terror may be created. Take a religious Hindu Brahmin and force such a person to eat meat—that could create terror . . .

Shilpa: I would like to discuss prejudice. Coming from Bombay, I can say over time it really has changed people. I'm interested in structure and politics, how this change happens and the fact that it's often invisible.

Mahzarin: That's the crucial phrase—not visible . . .

Shilpa: How much of our thinking is unconscious? . . .

Mahzarin: You can't offer any meaningful number. But Eric Kandel who studies the biology of learning and memory and won a Nobel Prize a few years ago says between 80 and 90% of our mental life is unconscious. A psychologist at Yale, John Bargh, says close to 100% is unconscious.

Shilpa: But human actions come from the conscious.

Mahzarin: In one sense, not really . . .

The questioning turned more immediate. A couple of weeks before Shilpa's last trip to Cambridge, she had locked her apartment door as Mumbai was thrown under curfew and the noise of gunfire spilled into the streets. Emails had passed back and forth between Paris and Mumbai as we'd reacted to the headlines. Shilpa had responded with clear brief thoughts. Yes, the Taj was in the hands of terrorists. Had they come from

Pakistan? The city had been brought to its knees by a handful of young, trained, fearless terrorists. Apparently they'd been drugged. She hoped to be able to leave the city by Sunday. We were all to meet in the United States early the next week. This latest experience, and many others she'd lived through over the previous years, sharpened Shilpa's questioning:

Shilpa: Are there certain emotions that come to us faster than others? Is hate an easier emotion than love or compassion?

Mahzarin: It almost seems that way. It almost seems that the more noble emotions are ones that we have to strive for whereas the base, negative ones just come to us. I'm not sure about hate but there are certain negative emotions that are more powerful than the positive ones in the sense that they get elicited rapidly. And again it is from a survival point of view. If I don't feel happy about something I should be happy about, it's not going to cost me my life, but if I don't feel fear I will not run away.

Shilpa: What's the opposite of fear.

Mahzarin: There isn't one! Approach?

Shilpa: Concerning fear, when I asked you if fear was genetic, you said that yes, it probably was.

Mahzarin: I would say that the ability to feel fear is built into us. But what evokes fear is largely learned through experience.

Shilpa: Would you say terrorism is genetic? Acting in order to cause pain? If fear is genetic, the opposite would be terror, causing pain. The next question is a bit more problematic. If we were to think that terror is genetic, if fear was genetic, what about this crazy but popular belief in India and all over the world that Arab Muslims are more prone to causing violence than others?

Mahzarin: We know this is simply not the case. If you look at the history of human beings on every continent, the continent that has the longest history of extreme violence is probably Europe. But modern Europe is among the least violent continents. So if it was genetic, how come it used to be that way but isn't so now? If there are differences by continent or culture, they are small enough to be less interesting in comparison to the fact that all human beings seem to be similar to each other.

For weeks prior to her show Shilpa worked at Le Laboratoire to finish the details, with the help of our technical staff. It led to two new works of art. Shilpa's *Singing Cloud* had four thousand microphones hanging from the ceiling in the form of a kidney-shaped cloud, some with speakers hidden inside so that you heard the singing voice of the artist amid sounds that ranged from the rustling of a stream to the stamping of an immigration seal.

The exhibit included three previous works of the artist, plus a second original installation, a flap board of the kind commonly seen in train stations. "I'm interested in information that you can see and relocate from another angle," she explained in her first conversation with Mahzarin, when she described her thinking behind this work. "'Thou shall not kill' versus 'Thou shall not kill unless you really want to,' playing with meaning. I'm interested in media and how it can twist things and how meaning can change as it suits us." The flap board did this with a mechanical simplicity that disguised the intentions of a programmer, who sometimes changed the letters predictably, as if this flap board really were there to serve you, but at most times for reasons you scrambled to guess at. Shilpa's flap board had twenty-nine letters that assumed the names of literal destinations. At one point you seemed to be on a historical trip between India and Pakistan just after the two states formed—this being Shilpa's point of departure for the flap board experience. The flap board also provided information gathered through the course of the Cloud Foundation experiments on fear, and quotes from an interview we'd arranged for Shilpa along the way with the MIT linguist Noam Chomsky.

Spectators probably saw the hand of Dick Gross in Ryoji's exhibition more obviously than they did the hand of Mahzarin Banaji in Shilpa's, but Mahzarin's own reaction was strong. She arrived in Paris the week after we opened and participated in an evening of public conversation around the work. "I never thought I'd live to see what the unconscious mind actually looks like!" Mahzarin said in her first words to the public. You could walk into the "médiathèque" next to the exhibition gallery and take the IAT yourself on one of two computers we made available to visitors. And almost everyone who visited Shilpa's exhibit did just this—two minutes to get a glimpse of their own unconscious reactions to race, religion, skin color, and gender.

"The installations . . . are not easy to access," wrote Sonia Desprez for the February 20, 2009, *Journal de Dimanche* in Paris. "They are more poetic than didactic, incorporate sounds and images to speak directly to our unconscious minds: a cloud that sings, a multitude of microphones that have become crazy, an airport sign that tells of tragedies, numbers of dead, or words, such as Hindu or Muslim. Even more interesting," according to Desprez, was the "médiathèque of the exhibition space: we can learn here of the research underlying the exhibit, and take the Implicit Association Test. But attention! To learn we're not exactly who we imagine to be, isn't easy to digest."

The creative process of Le Laboratoire came to life with the Shilpa Gupta exhibition. One way you saw this was through the exhibition itself. Another way was through the experiment MuseTrek, which brought Shilpa's experiment to the Louvre, a couple of hundred meters away from the exhibition space. With a handheld interactive guide, visitors explored ideas related to Shilpa's experiment within the Louvre's permanent collection. High school students in Boston and in Singapore prepared MuseTrek's creative content. This project also catalyzed Le Laboratoire's commitment to commercial innovation, as Shilpa's experiment at the Louvre was the first step toward a startup company. Later that year Shilpa's exhibition traveled to the Louisiana Museum of Modern Art outside Copenhagen, and a year after this the Louisiana Museum purchased *Singing Cloud* and the flap board pieces. Here, then, was a cultural innovation that came out of this sixth Laboratoire experiment.

With Shilpa Gupta's exhibition, we began to clearly see the value in, and even the possibility of, quick experimental translation of ideas between multiple artscience labs.

$$\Longrightarrow\Longleftarrow$$

At the Science Gallery in Dublin and at the Wellcome Collection in London, exhibitions challenge visitors to see works of art and design as objects in evolution too, though more commonly through a perceived topicality of the exhibition. Art and medical science meet, for instance, in the work of John Isaacs in the Wellcome Collection's permanent exhibit on *Medicine Now*, where a large humanoid figure of wax, expanding foam, and oil paint explores the issue of obesity. Isaacs describes *I can't help the way I feel* (2003) as an exploration of the "representational possibility of the emotional landscape of the body becoming manifest in its surface." Looking at Isaacs' work, we can see a major medical issue of our time but also question how differently this will appear to us many years from now when the surprise of the form, so associated with the particular timeliness of Isaacs' question, wears away.

An exhibition on Artbots—robots as art—at the Science Gallery showed the latest research in robotics, with artist John Healy and computer scientist Peter Redmond collaborating on installations that would clearly lose their edginess in another five years, but might—if the exhibition could be viewed as a window into the world of robotics research—evolve in ways hinted at by the installations themselves, eventually swarming around human life in fascinating if frightening ways.

Annie Cattrell's *Sense* (2001–2003), in Wellcome's permanent collection, captures, through rapid-prototyping sculpture based on MRIs, constructions of the human brain as it responds to touch, smell, sight, hearing, and taste. We don't know quite what to make of these MRI images, any more than do scientists today. That sensations stimulate different parts of the brain is a relatively recent insight. MRI provides only a crude picture of this process, and neuroscientists continue to search for more accurate and meaningful measures. Staring at

Annie Cattrell's sculptures, we feel this push of knowledge—we see ourselves at a precarious point in a long, unending path of idea translation.

A project called *Seed Dating* at the Science Gallery aims to spark new creative collaborations by inviting the public for a drink in a fast-paced game of musical chairs, literally. Scientists, artists, entrepreneurs—anyone who comes—move from chair to chair while music plays. The goal is to brainstorm an idea of creative collaboration in a short period of time. At the Wellcome Collection, weekly lectures and events invite the public to participate in debate and sometimes in a breathtaking creative process. In May 2009, Paul Grundy of the Wessex Neurological Centre at Southampton University Hospitals performed brain surgery that was streamed live to an audience at the Wellcome Collection. Visitors were invited to ask questions of Grundy and to see the brain through the eyes of a neurosurgeon.

Audiences in these and many other contemporary art-science exhibitions participate in a collective inquiry, where the ability to hypothesize and dream is just as valuable and essential as the ability to analyze and deduce.

The contemporary art installation of French artist Fabrice Hyber, following his collaboration with MIT scientist Robert Langer, for the fall 2007 opening exhibition of Le Laboratoire. Imagining the possibility of sharing the experience of a stem cell that transforms into a neuron, Fabrice made several large hourglass objects out of different materials, including the biodegradable polymers Langer used in his research to attract stem cells or deliver them into the body. In one of these hourglass installations, visible at the foreground of the photograph, a visitor could, in principle, climb up on a ladder and slide through an elastic sleeve surrounded by small white beads that climbed upward as the visitor fell downward. This effect produced a maximum pressure around the body at the midsection of the sleeve, with beads and human body moving up and down in opposite directions, as if the visitor were a sand particle in an hourglass, or a stem cell undergoing the otherwise unimaginable process of cellular division. Other installations and paintings provided different hourglass figures, many with rotting fruit, which gave another way of thinking about the cellular division by which stem cells produced neurons. (© Marc Domage)

The culinary art installation of French chef Thierry Marx, following his collaboration with Ecole Physique Chimie colloid physicist Jérôme Bibette, for the spring 2008 exhibition at Le Laboratoire. Visitors received bento boxes with caviar-like eggs of flavor designed by Thierry, using Jérôme's calcium alginate membranes, which were extremely thin in order to leave very little calcium alginate inside the eggs, as this would harm the taste. Visitors looked on as servers prepared the bento boxes, arranging the tiny beads of flavor inside pears, or atop salmon. Visitors took their boxes to specially designed tables, where they ate unusual dishes, like tomato anchovy salad or celery root mousseline, while watching films by the artist Mathilde de l'Ecotais projected from the ceiling onto the tables, their plates, their hands. The films captured the process of creating these and other experimental meals. Later, visitors walked over to the Whif Bar, where they had a coffee and a whiff of chocolate, and watched a little machine making one egg of flavor after the next. (© Marc Domage)

The photographic installation of American photographer James Nachtwey, following his collaboration with Harvard Medical School doctor and scientist Anne Goldfeld, for the winter 2008 exhibition at Le Laboratoire. James presented large black-and-white prints of images he had taken of AIDS and tuberculosis patients who had access to clinics and received encouragement by scientists, especially Anne Goldfeld, professor of medicine at the Medical School and cofounder of a leading clinic in Cambodia. The images often bore witness to terrible pain and agony, but invariably they also shared the silent battle waged by health workers in out-of-sight places in a war that drugs, available in wealthier locations, might win in a better-managed world. This experiment, which provided the context for a major global health meeting in Paris and another sponsored by the Bill & Melinda Gates Foundation in Bangkok, explored how an artist might provoke scientists to think differently about their work in labs, with students, and in the field. (© Marc Domage)

The contemporary art installation of Japanese digital composer Ryoji Ikeda, following his collaboration with Harvard number theorist Benedict Gross, for the fall 2008 exhibition at Le Laboratoire. In a dark, silent exhibition space, two large rectangular prints, suspended over the floor at knee-level, presented millions of digits, in one case black digits on white, and in the other white digits on black. Visitors made out these digits with the help of a magnifying glass. One of these two prints presented a Mersenne prime, a number generated by the simple formulaic expression $2^p - 1$, where p is a prime number. The integers 2, 3, 5, 7, 13, 17, 19, 31, 61, and 89 are all Mersenne primes, but at larger numbers these special primes quickly become very rare. Prior to this exhibit a Mersenne prime number had never been presented in such detail—this one contains over seven million digits. If even one of these digits was removed, this number would no longer be a Mersenne prime. Next to this print in the exhibit was a print displaying an irrational number, meaning a real number that cannot be represented as repeating or terminal decimals. Since real numbers are uncountable and rational numbers are countable, it follows that most real numbers are irrational; therefore, unlike the case of Mersenne primes, removing one digit from an irrational number is likely to yield a second irrational number. Ryoji conceived of these images as representing the sublime and the beautiful, a visual art reflection on the meaning of infinity. (© Marc Domage)

The contemporary art exhibition of Indian artist Shilpa Gupta, following her collaboration with Harvard psychologist and neuroscientist Mahzarin Banaji, for the
winter 2009 exhibition at Le Laboratoire. In the center of the exhibition space
hung a kind of cloud of microphones, strung together around a metal frame,
many of the microphones having hidden speakers from which emanated voices
and sounds—singing, whispering, talking, all generated by Shilpa, mixed with
sounds of train whistles, rustling water, immigration stamps slamming against a
table. Visitors listened to the words and sounds as they moved counterclockwise
around the cloud, like a circular wave. In the background appeared a flap board,
of the kind that once displayed destinations in train stations. Here, Shilpa provided a kind of train trip of terror, which began with the train ride from India to
Pakistan following the division of the two countries. The purpose of the trip was
to carry across the new borders the dead who had perished in fighting. Shilpa
explored here the meaning of terror and its origin in the unconscious mind, with
the cloud a kind of visible consciousness of the majority of the world's population who lack a clear voice in the operation of society. (© Marc Domage)

The architecture and design exhibition of French architect François Roche, following his collaboration with French mathematician François Jouve, for the winter 2010 exhibition at Le Laboratoire. François Roche divided up the exhibition space with flexible semi-transparent walls that guided visitors through a kind of research lab maze. At each of five lab stations, visitors could participate in an ongoing experiment between the architect and many scientists and engineers to create an architecture formed and realized entirely in response to biometric data drawn from the bodies of those for whom the architecture would be designed. In the first station, in the foreground of the picture, visitors placed their hands on a glove-like sensor and responded to the questions of an actor, who assumed the role of a nurse administering a test. A screen displayed the movement of a robot that responded to the biological rhythms of the visitor and finally produced a numerical score, mimicking the analysis of data necessary to program the robot that would build a local habitat for the visitor. Next came the robot itself, a prototype of a polymer cement-extruding machine that could build the structure by climbing up and down, day and night. This robot was shown in simulated action against one of the flexible walls. In another lab station the visitor could see models of the architecture this biometric information produced—structures that resembled coral reefs or the shantytowns of Mexico City. The last two lab stations showed larger structures and the scientific and visual data underlying the project. (© Phase One Photography)

The design exhibition of Le Whaf by French designer Marc Bretillot, following his collaboration with the author, for the fall 2009 exhibition at Le Laboratoire. Visitors could fill a glass with a cloud of flavor, say a classic martini, and sip the cloud through a specially designed straw. To produce the cloud, piezoelectric crystals produced ultrasound waves, generating cavitation in the liquid. (© Bruno Cogez)

The design exhibition of Bel-Air by French designer Mathieu Lehanneur, following his collaboration with the author, for the fall 2007 exhibition at Le Laboratoire. Five unique designs were distributed around a room whose floor was covered with synthetic carpet, which was known to produce invisible toxic gases such as formaldehyde and benzene. These gases, represented on the walls and columns of the space with charcoal drawings, were drawn into Bel-Air by a fan, which pulled air past a plant, through the plant's soil, and back into the room over a water bath. Bel-Air filtered the air and cleaned out the gas, which it deposited on the leaves, in the soil, and onto bacteria within the soil. Some degree of metabolism by the plant removed the gas and degraded it. During the exhibition, experiments recorded the efficacy of the plant filter, and these results were produced on a white board against the wall. The commercial plant filter Andrea, which began to sell two years later, was derived from this original design, but at a smaller size and with simplified materials and function. (© Marc Domage)

Whiffing chocolate on the Majestic Beach in Cannes during the 2009 Cannes Film Festival. Le Whif was distributed to visitors during the traditional Majestic Beach lunch—a whiff of chocolate following a meal, or, in this case, while watching boats on the water. This was the first experiment organized outside Le Laboratoire, a few weeks after the launch of the first Le Whif prototoype. The enthusiasm of Cannes festival-goers played a key role in encouraging the LaboGroup to continue development of Le Whif, which returned to Cannes the next year with a new design, and the new experience of whiffed coffee. (© LaboGroup)

The LaboBrain office of Le Laboratoire, designed in 2008 by French designer Mathieu Lehanneur. Green plants grew up through a metal grate in the floor. Walking over the green floor led a visitor past desks where leaders of the Labo-Group and ArtScience Labs sat, to a kind of cave formed by a giant white board made of fiberglass and on which notes, designs, and other ideas could be drawn and erased. In front of the white board, various forms, in the shape of large champagne corks or screws, provided visitors with seating or tables. A large beanbag, designed with a geodesic form, became the seat or couch on which a member of Le Laboratoire, frequently the author, sat to reflect and discuss. A plasma screen swiveled on a column in the window and became animation for the street, showing a film that captured the previous years' programming, or, when needed, served as a screen on which meeting notes could be projected. (© Klane Fabien Thouvenin)

═5═

INNOVATION

Following exhibition in Paris, works of art or design often spark further experimentation, reaching progressively larger publics through shows in galleries and museums around the world. Design exhibitions occasionally reveal commercial promise, too. And when they do, novel works of design may emerge through rapid prototyping, evolving through experimental development and early-stage sales in our translational change lab, the LaboGroup.

The oddness of these pre-commercial ideas, their surprising associations with new ways of living, make them less bets on what is relevant today than guesses of what will be relevant tomorrow, and therefore fascinating grounds for experimental inquiry. The public participates in the innovation process almost as lab members—as if, like readers of scientific publications, their mandate is to question, critique, and carry ideas further. This phase of cultural incubation helps lower development barriers to ideas, which, if regarded from the outset on the basis of their commercial promise, might seem too high-risk.

While the possibility of successful commercialization can drive ideas through the narrow end of the idea funnel, if commercial potential is taken too seriously early on, it can also cloud lab vision. Assigning a commercial value to experimentation may invert the natural value of time, by hurrying ideas along toward a potential payoff, the quicker the better. It can rip the created thing away from the creator, who often tries to expand time, and into the hands of the developer, who tries to shrink it, at just the moment when the created thing itself —if it is truly innovative—is in almost constant need of re-creation.

If creators know that by coming to the lab they are making a bet that what they dream about will pay off commercially, they tend to dream less. But if commercial success remains the welcome surprise and not the goal, cultural incubation within artscience labs should lead to especially innovative outcomes, which assume, through cultural incubation, increasingly mature forms before the public eye, as I describe in this chapter.

Cultural incubation pushes creators into a frank, refreshing, often provocative public dialog around the creative process. Beyond their role as buyers, members of the public become intimate participants in the how and why of the creative process. They are experimenters in the lab to the degree that their opinions, expressed through reactions to the exhibition, purchases of product prototypes, and commentary in the press, shape the future experimental process.

≡≡

While studying at the Ecole Normale Supérieure de Création Industrielle (ENSCI), Mathieu Lehanneur had designed several clever tools for delivering drugs like cold pills and asthma medications. A few years had passed since then, and while his inventive designs had entered the permanent collections of MoMA in New York and San Francisco, they had not made much of a ripple with industry. They had also not been patented, a common practice among designers in France, where the rights to artistic works remain more inalienably with creators than they do in the United States.

Mathieu had come to the Paris lab's attention a year before its opening through a friend of Adjunct Director Olivier Borgeaud. Considered among the remarkable young French designers and with some experience working alongside scientists, Mathieu proved affable, direct, and anxious to learn. During our first conversation, he described his recent entry into the medical science field as you might talk about happening onto a big blue lake after an interminable walk through dense woods. His mind worked more freely outside the normal constraints of contemporary design. He felt more creative approaching problems he'd not been trained to understand.

Mathieu showed images of medical product concepts he'd thought up, including an asthma inhaler that inflated with air overnight as a child slept, so that, when the child awakened, he was actually delivering a service to the inhaler, by emptying it, in order to do a service to himself, by filling his own lungs with the excess air. He also showed the design of a stirring rod that was a drug, so that a patient created a ritual every time he digested it, working the glass of water with the stirring rod, getting ready to be treated or cured.

Industrial design is often an effort to better adapt useful scientific inventions to human intuition and practice. With Mathieu it was something quite different. His design *was* invention. I asked if he would be interested in exploring another open field, one requiring that he dig more deeply into the soil of scientific research and learn what was hard to know. In order to create new industrial designs at the frontier where scientists were creating new science, he would have to make discoveries, be surprised, and change his idea path more than once. Mathieu liked this proposal, and welcomed the possibility of being surprised.

The topic Mathieu eventually identified was air purification by plants, a recognized natural phenomenon but still poorly understood by scientists. Researchers had known for more than thirty years that plants absorb and metabolize toxic gases in the air. But how they did this and to what extent they significantly cleaned the air remained unclear. Did plants actually lower the concentrations of toxic gases in home and office settings, where windows, doors, and other sources of air circulation led to rapid air exchange with the external environment? Not knowing answers to these questions, researchers had stopped short of figuring out how to efficiently, effectively, and practically take advantage of this potentially beneficial phenomenon.

We spent the year 2006–2007 collaborating on a new invention to clean indoor air by moving it through the leaves and soil of absorptive plants. The design that resulted, Bel-Air, eventually reached the commercial market as Andrea (named after Mathieu's two-year-old son). Prior to the Bel-Air exhibition in the fall of 2007, we patented our design, not because we saw an obvious commercial future for the product but because we wanted to leave the door open for surprise. To our delight, retailers and distributors approached us during the exhibition at Le Laboratoire. Did the filter actually work? Our data, taken during the exhibition, indicated that the design filtered air about a thousand percent faster than absorptive plants could do alone. Could the product be mass-produced at a reasonable price? We had not yet looked into it. At its current cost of

15,000 euros each, we were hardly surprised that our prototype attracted zero customer interest.

When the exhibition came to an end, the best measure we had of the innovative value of this first experiment was the many articles that had appeared in the press and the positive reactions of visitors to the lab. The design traveled the following spring to an exhibition at MoMA in New York and won a Popular Science Invention of the Year award, which added to our sense that we were on an innovative trajectory and encouraged us to continue experimentation.

This development moved the idea outside the scope and competence of Le Laboratoire, and so we created LaboGroup as a French commercialization company led by the entrepreneur José Sanchez. The experiment continued under José's guidance. We needed to redesign the filter to lower the cost by a factor of about a hundred, find a manufacturer who could guarantee global supply, and explore among the tens of distributors who approached us during the exhibition whatever credible channels of sale might exist that would make development of Bel-Air worth everyone's time and investment.

Were we up to all this? It was easy to say we were not. The LaboGroup consisted of just a few people, and nobody had professional experience precisely pertinent to the manufacture of anything, let alone plant filters. Mathieu wondered if we might not do better to have Bel-Air manufactured by one of the large companies that had expressed early interest in taking over the project. But the risk was too great, and the reward too small, to hand over the rights to the design before we'd shown that it could be manufactured and eventually sold. By moving the plant filter idea from an artistic design to a commercial product ourselves, we were bound to learn things, not all favorable. As an idea matures, it can appear a thousand times innocent, flawed, too far from its target to have any hope of getting there. Without an intimate stake in the survival of an idea, in the realization of a dream, an idea translator might abandon it. For this reason at least, creators hesitate to turn over idea development to others too quickly.

Mathieu, José, and I discussed all this during many meals and drinks. Meanwhile, we opened up our first serious conversation with a potential partner, Monceau Fleurs, the leading seller of plants and flowers in Europe. Monceau Fleurs was founded in 1965 by Laurent Merlino with a first store at a corner of the boulevard Malesherbes near Paris's Parc Monceau. The model of a well-positioned shop that sold fresh flowers at competitive prices soon dominated the French florist market, and by our time, under the leadership of Merlino's grandson, Laurent Amar, Monceau Fleurs was expanding outside plant sales.

Laurent, a smart and admired entrepreneur with a nose for innovation, invited the three of us over to the Monceau Fleurs headquarters. Yes, he liked Bel-Air, believed it matched the future direction of the company—flowers for personal well-being—but he doubted we knew enough to reliably manufacture it. By spring 2008 we came back to him with drawings of a first commercial prototype, which we called Andrea—a trade name we figured we could protect in the marketplace. In the two months between our first meeting and this one, José had spoken with a friend of his who led production for the manufacturing company Holinail in Shenzen, China. The pair had worked together before and trusted each other enough to start a speculative collaboration. Mathieu had looked over the industrial design and made some adjustments to the drawings, and then we turned them over to Laurent.

He still didn't believe, and provided plenty of good reasons not to. We had done no marketing research, the retail price of 500 euros was obviously too high, and even with drawings from our Chinese manufacturer, the board of Monceau Fleurs could never be convinced that our scrappy, inexperienced team would effectively manage production. Mathieu and I passed some more evenings together. Yes, we were surely naive, but the process was fun, the team was invigorated, and if I was willing to bet time and money on it, so would he. By the end of summer 2008 we had our first prototypes. We opened the fall season at Le Laboratoire by announcing that Andrea would be commercialized within the year (this turned out to be wrong), even if we had no idea who exactly would sell it.

Having the prototype helped us open up a partnering conversation with the French company Nature et Découvertes, heir to the old Nature Company in the United States. Representatives of Nature et Découvertes expressed interest in the product but said we needed to bring the price down. I pointed out to Mathieu that we could eliminate one of the two fans without altering the efficiency of the filter. José and Mathieu redesigned Andrea and produced a second prototype, which arrived in Paris in mid-December 2008. Another meeting was arranged with representatives of Nature et Découvertes, and this time they said yes, they wished to sell it. We had the seed of our first commercial distribution deal!

From here to commercial launch in the fall of 2009, we placed a series of escalating bets on a growing LaboGroup team. Tom Hadfield, who started a company in England when he was thirteen and sold it to ESPN while he was still a teenager, had come to my attention when he took my Idea Translation class in the fall of 2007. He joined in December 2007 to help craft our first business plan for Andrea. He also started conversations with the many distributors and retailers who had contacted us during the Bel-Air exhibition, leading to tentative agreements with the U.S. catalog retailers Frontgate and Hammacher Schlemmer. In spring 2008 Tom recruited an experienced businessman to head up sales and distribution in the United States, while Tom moved with his wife, Kristen, to Paris to help build up the commercial artscience lab. By summer 2009 Manuella Passard had joined to lead our manufacturing and production team. We signed new deals with Galeries Lafayette, Monceau Fleurs, BHV, and Amazon.

A new battle began. The product had no obvious precedent on the market. Conventional filters generally processed large quantities of air in relatively short periods of time and quickly cleaned out particulates. Andrea functioned more like a plant, filtering air slowly and capturing volatile gases. Was Andrea just an unusually smart flower vase, or was it truly a new, better kind of air filter? Educating distributors, retailers, and finally consumers became a major preoccupation.

Product started to ship in October 2009, and sales began in French stores by November. In the following weeks our staffers ran in and out of stores all around Paris, then throughout France, to see how Andrea was being displayed, to talk with managers, to learn if the product would sell, and why, and how. For Le Laboratoire, this marketing research was just a continuation of the experimental, educational, translational artscience process that had started in that first conversation with Mathieu. It was also our first experience of the cultural incubation process outside the walls of Le Laboratoire. We had started in an exhibition space. From there, we had skipped the diligence and deliberation that development of a plant filter would have demanded in a more traditional corporate environment. We brought Andrea into Monceau Fleurs with a spirit of experimentation, in about the same way that we sent works of art, with some adjustments, to larger exhibition spaces in New York, Brussels, Hong Kong, and other cities. The downside to this way of proceeding was that potential investors or partners could not see the commercial wisdom of our process. But this process kept us creatively engaged, and Mathieu grew even more passionate about his idea, drawing up new plant filter designs and stopping by my office to brainstorm on practically a weekly basis.

On one of these visits, he heard about a new project I had started with students on an idea related to water transport in arid low-income parts of the world. Did it interest him? Yes. And with that response, Mathieu started a second collaboration that led to a novel way of carrying water, which we eventually called the Pumpkin. The story of this idea is a subject of Chapter 6.

The Mathieu Lehanneur experiment didn't, of course, start in the fall of 2006 with any of these developments in mind. The original project had aimed to explore how to make plants more intelligent. Here was its justification—an original idea that would lead to a new design, a public exhibition. Idea translation inside the artscience lab environment had carried the original hypothesis (that we actually knew how to make plants "smarter," whatever that meant) to the production of a novel plant filter that actually sold in stores, and now to reflections on a novel way to transport water in Africa.

⇉⇇

Thierry Marx was helping transform how we enjoy the purely aesthetic realm of eating. Each year, in the town of Pauillac, north of Bordeaux, within the chateau of Cordeillan-Bages, he created hundreds of new ways to prepare, visualize, and consume familiar foods. By 2007 the reputation of his restaurant drew comparisons with the two top experimental restaurants in the world, Ferran Adrià's El Bulli outside Barcelona and Heston Blumenthal's The Fat Duck north of London.

That same year we had a chance to involve Thierry in an experiment at Le Laboratoire with the colloid physicist Jérôme Bibette. To explore how a chef became an exhibiting artist, we traveled down to his restaurant in July. The conversation swirled that day around wrapping flavor in particularly thin membranes. Having looked into the idea of inhaled aerosols for delivering drugs and vaccines, I brought up the idea of breathing these colloids into your mouth. Later in the fall I shared that notion with students at Harvard University. They would need to make the food particles small enough to get into the air, and large enough to avoid entry into the lungs under all conditions of breathing. We knew this much. But what did inhaling food mean? Would there be pleasure in it? After a semester of reflection, brainstorming, and quite a bit of coughing (even after designing the particles with a size to avoid the lungs, we discovered that, no matter how we breathed through straw-like inhalers, the particles flew to the back of the throat) I put a piece of tape over the paper cylinder my students had prepared to inhale things like carrot powder. The coughing stopped. And here we had the first prototype of the food inhaler we called Le Whif.

The LaboGroup was just then being formed, and José Sanchez decided we could manufacture Le Whif in the Chinese factory where we planned to make the plant filter. Six months after the idea had come up in Thierry's restaurant, we manufactured the first Whifs in China. They arrived in Paris a few days before Thierry and Jérôme's culinary art exhibition at Le Laboratoire, in March 2008. We exhibited our fledgling product over a Whif Bar imagined by Caroline Naphegyi and cosponsored by the Nestlé division Nespresso. Nespresso offered each visitor a free coffee, and we included a little brown object that looked like a tube of lipstick, by which you whiffed chocolate into your mouth.

To be perfectly frank, Le Whif didn't work well. Chocolate powder fell out if you inclined one end above the other, and you almost invariably coughed when you inhaled for the first time. But this was a lab, and we were testing a new idea—a new way of experiencing food! Two Harvard students, Trevor Martin and Larissa Zhou, had flown to Paris for the event. Jonathan Kamler, who had graduated the year before, having led the previous semester's student whiffing project, was also in Paris to work full-time for the LaboGroup. These three kept the chocolate inhalers full of chocolate powder as several hundred opening night guests tried it out.

It turned out to be fortuitous that, just weeks before the exhibition, the French government had outlawed cigarettes in cafés. This had outraged many French café-goers. Le Whif seemed a kind of inventive response. The traditional sip of coffee, bit of chocolate, and smoke that properly ended a French meal became, in this new anti-cigarette era, *sipped coffee and smoked chocolate.* Our guests had a ball with it. They invariably held Le Whif between their fingers as if it were a cigarette, and kept it long after the tube was empty, chatting, appreciating a novel social experience (which became, in the hands of my three little boys, something of a slightly illicit thrill).

No, this wasn't a commercial product, and nobody pretended it was. Nobody, that is, but the LaboGroup team. Why? Because the team was having fun. The hoped-for outcome of this first experiment had been observed in the public reaction to the exhibition, and now there were more experiments to be done. And, besides, if Le Whif did manage to become a product one day, magical revenues would appear. The team needed to hope for this income stream. The risk of running out of money was too palpable.

True, whiffing was even more far-fetched than filtering the air with plants. However, being far-fetched made the idea plausible as the preoccupation of an art lab, and commended the idea as a valid creative process even as it cast doubt on the eventual outcome of a valuable commercial product.

Clearly there were things to improve. The design needed to prevent all the chocolate from spilling out as you moved Le Whif around after filling it and before inhaling. But the product needed to remain simple; especially, it needed to avoid the unattractive trappings of pharmaceutical products. Le Whif needed to reliably deliver enough chocolate to satisfy taste but not so much as to fill your mouth with dry powder or provoke a spasm of coughing. The LaboGroup launched a second version of Le Whif in the fall of 2008 with the opening of the new LaboShop. The public could come inside and enjoy a whiff of chocolate with a cup of espresso. Whiffed chocolate came in mint chocolate, raspberry chocolate, and pure chocolate flavors. We invited the public to opine on the result and help us design a fully commercial chocolate inhaler that we would launch within the year.

We said that whiffing was a new way of eating, and proposed Le Whif as a kind of inhaled fork or spoon—and we believed it. Until now, nobody had reliably put food in the mouth through breathing. First there were the hands, then chopsticks, then forks and spoons, and now Le Whif. Yes, we were starting with chocolate, but you could inhale many other things, too, as we did during private evenings of experimentation in the FoodLab—with cheese, mushroom, exotic teas.

From the fall of 2008 into the winter of 2009, our local LaboShop clientele included young professionals, kids, and an occasional celebrity, like the French actress Isabelle Adjani, who would sneak in and out, sunglasses donned, enjoying a private pleasure. This eclectic group, probably fewer than a hundred people, returned regularly to the LaboShop, defining for us the commercial potential of Le Whif and rarely leaving before posing the same impatient question: When could they actually buy Le Whif and take it home? The LaboShop team would, when asked this question, point to the upcoming product launch, whose date we kept moving out further into the future as we tried to improve the design of Le Whif, now a third time, and work out a business model in the absence of a huge demand.

The team geared up for the commercial launch over regular late-evening drinks at the neighborhood café. Production costs needed to be lowered, simplicity improved, and more chocolate added, among other things.

With so much happening then at Le Laboratoire—the global financial crisis having fully settled down on us—our product development was spotty and informal. We did the best we could with limited time and resources. After many delays, the date of the launch was set for April 29, 2009. There would be a world tour to bring Le Whif to major cities and accustom people to the notion of breathing chocolate. We put together communication materials and planned a launch that resembled the opening of an art exhibition, which was the only kind of opening we knew. Our message was philosophical. Breathe chocolate and experience food as an artistic act. If chocolate failed, we had other ideas—inhaled spices, inhaled steak, inhaled coffee. Thierry Marx began to think about it all.

Four weeks before the scheduled launch in Paris, something very surprising happened. In his first months working for us, when he was still living in Cambridge, Tom Hadfield helped us put the business plan of LaboGroup in shape. In the midst of this he sent a note in early April saying that he was about to start a buzz campaign by Internet. Distracted by the challenges of the launch, we didn't take special notice. Tom's note probably arrived on a Thursday. The campaign was to begin the next day, he wrote.

On Saturday morning Tom reported that some blogging had started and the traffic on the Internet site had doubled. We received another note from Tom on Sunday. Internet traffic had doubled again. Similar messages came on Monday and

Tuesday. By then Internet orders for Le Whif were flooding in. Several major blog sites picked up the story midweek, and on Wednesday the *New York Post,* the *Chicago Tribune,* and the *Boston Globe* wanted to do interviews. "The world has been waiting for breathable chocolate!" Tom wrote ecstatically in one of the many emails now zipping across the ocean. By the end of the week the *Today Show, Good Morning America,* and *CBS Morning News* had asked for the product to test. We had waves of orders and media requests from England, Germany, Italy, Spain, South Africa, India, Thailand, Japan, Poland, and other countries.

The world had awakened to the idea that we had a new, surprising commercial product. But we actually didn't. We were unprepared. The day before the April 29 launch I was in Washington, D.C., to give a talk at the National Institutes of Health on work I was doing related to infectious diseases. On my way from the airport I received an email from Jonathan Kamler explaining that Le Whif had arrived, filled with chocolate and properly packaged. But, when he took it out of its packaging to test it, Jonathan discovered it didn't function properly. He couldn't even open it. Out of a hundred Whifs, perhaps thirty worked. In a lightning decision, we decided to hand pick the new product, throw out the defective ones, and launch the next day. Thankfully, the surprise of the product, the suddenness of international public reaction, and the bizarre atmosphere of the FoodLab, with Thierry Marx presiding over lunch, helped everyone ignore that day how unreliable this first product actually was.

The team was invited a few weeks later to the Cannes Film Festival to help animate the beachfront terrace café of the Majestic Hotel. For two weeks young women walked between café tables from noon through mid-afternoon. Shoulder straps held serving trays from which they offered free Whifs, like vendors selling hotdogs at a baseball game. Later that month the team traveled to Chicago for the All Candy Show. This went relatively smoothly, and by the end of the month we ran out of the first faulty stock—15,000 Whifs.

A new shipment arrived in July, and until October 2009 it performed mostly as we wished it to, particularly in the hands of those educated to use it or patient enough to learn how to use it even when it dusted your knees with chocolate, or when the chocolate flew into your mouth in one quick burst, or when the inhaler arrived empty, because of some glitch in the filling, packaging, and transporting process. While we received many disappointed notes from customers who received Le Whif and either did not understand it or experienced a malfunction in one of these ways, we also received notes from at least as many customers delighted by the product, who understood Le Whif even with all its youthful blemishes, and who remained hopeful about ordering more, particularly once we'd figured out the issues of manufacture and supply.

The experiment continued. In October we produced an even more reliable chocolate inhaler that began to sell in Lafayette Gourmet within the flagship Galeries Lafayette store. Helped by a brilliant chocolate expert, we completely changed the packaging and marketing of the product to better signal our commercial intentions while preparing the launch of a fully commercial Whif at the end of January 2010. This final product was launched at Davos, Switzerland, during the World Economic Forum, and later that spring in Dylan's Candy Bar in

New York City and other locations around the United States, such as the gourmet shop Cardullo's in Harvard Square. That same spring the product launched in England within the House of Fraser in London and in many other cities and towns around the United Kingdom.

Le Whif's commercial appearance mobilized the entire network of artscience labs. LaboGroup ran the business, Le Laboratoire curated the idea through exhibitions, The Laboratory at Harvard introduced the product during its opening in 2009, and the Idea Translation Lab at Cloud Place organized a few high school whiffing parties. Through sales, Le Whif would benefit all the artscience labs eventually; but this was not the reason why Le Whif went on exhibit in Paris or animated parties in the United States. The labs participated in this experiment mostly because Le Whif was a surprising idea, conceived with students at the intersection of aerosol science and culinary art, and, while fun, it also expressed something essential about what each lab did, or wished to do, with students, creators, and the public.

More than a commercial product, Le Whif carried the creative process outside lab borders. We wished the product to be understood in its original art-as-process context. This was signaled by the launch parties in Paris and, later, at the Cannes Film Festival, in Chicago, in Cambridge, and elsewhere. It was signaled by the early silver balloon packaging we used, and by the "airline tickets" we handed out to explain what we had in mind with the "world tour" promotion. Our sales approach reflected a lab sensibility; it did not reflect a reasoned analysis of the market.

Le Whif traversed the entire idea funnel. It started as a catalyst of education, soon became a catalyst of cultural exploration, and went on to be a catalyst of commercial sales revenue that helped keep our labs running. It also inspired new culinary art and science experiments, from whiffed coffee, which launched in the spring of 2010, to whiffed vitamins, scheduled to launch later within the year. And on the horizon was yet another design, Le Whaf, which I conceived with the French designer Marc Bretillot as a new way to "drink by breathing." This was a new form of food—a standing cloud of flavor that falls between a liquid and a gas, just as whiffed food fell somewhere between a solid and a gas.

⇒⇐

In the same month we pre-launched Le Whif, we began another experiment outside Le Laboratoire. For three Friday evenings in April 2009, visitors to the Louvre who walked past Bernardino Luini's early sixteenth-century *Adoration of the Magi* saw a member of the Laboratoire team sitting at a table in front of the painting and handing out iPhones, contributed by our partner France Telecom. On each iPhone was an application we called MuseTrek.

MuseTrek had emerged in the classroom as a new cell phone and Internet approach to cultural exploration that gave visitors the chance to explore museums and other cultural destinations about as easily as they surfed the World Wide Web. Before MuseTrek and other hand-held approaches to cultural exploration that began to appear around that time, museum media guides had mostly focused on associating useful information with museum objects and linking this information together into a guide that helped visitors navigate their way through a kind of cultural maze. Largely this information came from the curatorial staff, making the museum's approach to visitors resemble a university's approach to students. Spectators received information, assumed to be educational and possibly even entertaining, which had been filtered through educated minds. The goal of MuseTrek was in a sense to increase the exchange of ideas through the filter, so that the public associated the filter of cultural interpretation with their own processes of learning and judgment.

MuseTrek, as originally conceived by students in the Idea Translation class at Harvard in the fall of 2007, invited anyone in the world to upload stories or "treks" associated with cultural objects onto a website, and to download these treks while visiting the cultural venue onto a handheld device. These treks might be based on an artist's life, following the chronology of the artist's career; or they might be historical, grouping cultural objects by epoch; or they might be adventurous, inviting the creation of games or treasure hunts; or they might be just about anything that a visitor could imagine. They were simply a reason to come to the cultural venue. Possibly visitors came to see, learn, and understand an official view of culture provided by a curatorial staff, but there could be many other reasons, and visitors were invited to share them.

The information a visitor associated with a given museum object might mean nothing outside the context of the trek within which the information became embedded. What pulled a visitor from one object to the next was not so much the conventional facts—when it was made, why it was made, what purpose it served—as the place the object held in an adventure that the uploaded trek created within the visitor's mind. Muse-Trek did not eliminate the role of educated experts, and it did not preclude the existence of curatorial treks where each station or artwork on a trek led a visitor to meditate on factual, historical, or conventional interpretive information related to the artwork. The idea was, rather, that this kind of scholarly trek would coexist with all kinds of other treks.

During the first public experiment, at the Louvre, visitors to the museum could hand over an identity card and receive an iPhone. It opened with an image of Luini's *Adoration* and an invitation to explore the artist Shilpa Gupta's trek through the museum, or to choose another itinerary. If visitors chose Shilpa's trek, which related to her then-running exhibition at Le Laboratoire next door, they discovered a screen that displayed a small section of the *Adoration of the Magi* painting and a poem from the artist. This led to another small section of the painting, and eventually visitors would find themselves exploring this entire work of Renaissance Italian art from the perspective of a contemporary Indian artist interested in modern questions of prejudice and fear. Visitors looking for a different trek came upon a screen that listed several themes, including historical, comical, and personal narratives. On selecting a theme, they discovered treks that took them to other paintings in the gallery, which they visited, iPhone in hand, immersed in some new perspective.

The MuseTrek experiment at the Louvre, like that of Le Whif when it was first presented to the public as a prototype, or even Andrea when displayed in the LaboShop under the name of Bel-Air, did not pretend to meet a cultural and commercial need. Six months later MuseTrek would progress toward that end through further experimentation within the context of an independent startup company and an exhibition at Le Laboratoire in fall 2009, but in April of that year Muse-Trek was still at the transition point between a student project, emerging out of the idea translation labs at Harvard University and Cloud Place, and a commercial business.

Mishy Harman, a senior social studies concentrator at Harvard, Tarik Umar, a sophomore economics concentrator, and teaching fellow Aviva Presser had led the conception and development of this idea during the fall of 2007. A friend of Mishy's, Roee Gilron, then studying at Brandeis University, had joined the team in the winter of 2008. They agreed that cultural objects in museums were losing meaning with the new generation. Television screen-time had gone down, computer screen-time was up, and young people especially were finding cultural enrichment through interactive environments and dialog. But when they walked into a museum (or into a library, such as the Widener Library on the Harvard campus), they could be perplexed about where to start. Ideally, they needed a way to somehow surf for information by simply walking along, a way to enter into a drama of exploration by which they might make any of this culture their own. MuseTrek came out of the students' desire to remove bewilderment from the experience of visiting a cultural institution.

Building a company out of this vision, which lacked an obvious business model, involved the merger of several interests, each anchored in the value of the experimental process, nearly independently of where this process might lead. The founding students, especially Mishy Harman and Roee Gilron following their undergraduate studies, saw in the pursuit of a business an opportunity to learn how to start and develop a new technology company. They were passionately devoted to their idea, wherever it led. If a successful company resulted from their commitment of a year or two after school, they would continue with it and build early entrepreneurial careers, and if it did not, the experience would likely benefit them whatever they next did.

Leaders of the ArtScience Prize saw an interest as well. Designing treks and building a cultural guide business that suited the approach to culture of the young generation might be an exciting context for dream generation and development. If a commercial business emerged, revenues could support the charitable cause of the ArtScience Prize, and if it didn't, the process of experimentation in some of the world's major museums would prove educationally valuable anyway. The Art-Science Prize team decided to involve students in the project and to support its development. Leaders of Le Laboratoire also saw an opportunity in MuseTrek—another possibility for experimenting outside the walls of the exhibition space and partnering with cultural institutions in Paris. This experimentation might produce a commercial business, but even if it did not, it had a clear cultural value and commended investment in the idea.

The entrepreneur Bill Jacobson, cofounder of the venture capital incubator TechPoint Ventures, invested in MuseTrek and became a cofounder in late fall 2008. He was MuseTrek's first CEO. Bill saw in this idea the solution to a cultural need that many museums around the world wanted to meet. He also perceived an opportunity in the emerging crowd-sourcing business, with frontrunners like the company foursquare.

MuseTrek might bring this business model to cultural organizations. And while the business plan remained obscure, the combination of student and charitable interest in the early-stage company, with all the resources of time, money, and access to the cultural world that these interests carried with them, lowered the risk sufficiently for Bill to invest his own time and money in the opportunity.

MuseTrek grew as a fledgling business from the fall of 2008 through 2010 with partners including several museums in Paris and a growing list of museums and galleries internationally. Early funding came from partnerships and European Union grants, thanks to Noemie Tassel, who headed the company's effort in Europe, working from Le Laboratoire.

Through the creativity of Cashman Arbus, an experienced programmer from MIT working out of Brazil, the technology moved closer to the crowd-sourcing vision, while the team's business model, brainstormed by Bill, Mishy, and Roee, finally focused on three objectives: to build deep relationships between new young audiences and cultural institutions; to engage these young audiences in such a vivid dialog that cultural institutions became part of a dynamic sociocultural exchange; and to expand these sociocultural networks to include other cultural venues, from public parks to restaurants, breaking down cultural boundaries and helping to ensure long-term cultural institutional growth.

≡≡

The original ideas described in this chapter were translated from conception to commercial realizations through the process of cultural incubation. Experimentation itself, almost independently of the commercial value this experimentation produced, generally justified investments of the time and passion of creators, of the labs within which these creators worked, and of the public that engaged in the experiments through exhibitions and early-stage purchases. In this process, hypothesis mattered more than utility, surprise more than functionality, and imagination more than commercial viability.

These kinds of ideas, and this process, are particularly suited to artscience labs, where experimentation around far-reaching hypothetical ideas aimed at beneficial social impact provides a combination of educational, cultural, humanitarian, and commercial value. If, on exiting the labs, ideas flourish in the marketplace as truly innovative products, it is surely related to the fact that for many years it didn't quite matter.

≡6≡

ALTRUISM

Self-interest can be a surprisingly social embrace. Even a miser, who worries about his money but knows that most of it will never serve him, is contemplating interests other than his own when he makes designs for his fortune long after he's dead. We speak of "our" neighborhood, "our" city, "our" country, "our" culture, "our" world—and in speaking this way we map out a set of interests that reach well beyond our own.

As Charles Sanders Peirce, the early twentieth-century logician, argues in his *Chance, Love and Logic,* logic anchors individual interest in a vast sea. We translate ideas—or, as Peirce puts it, we inquire—to remove doubt, which we run from in pursuit of the more satisfying state of belief. Through inquiry, we face up to our inferences. In Peirce's famous example, we enter a room of sacks randomly filled with black and white beans; we approach one, reach in, and pull out a white bean; and from this we infer that all the beans in the sack are white, which they may not be. Our chance of being wrong is equal to our chance of being right. If we're set on finding a sack of all-white beans, we reach back in and pull out a fistful of beans. If our possibilities (or dreams) are limited to this one sack, our chances of getting nothing but white beans are not so good. But if we lift our eyes and move on to other sacks, the odds improve. Peirce also brings up the example of the old Martingale betting strategy (more useful as a hypothetical idea than a practical approach to gambling), where each loss at a game of chance (imagine a coin flip) is followed by a bet that doubles the last bet made. Each time we lose, we bet again, twice the sum. Given unlimited funds, unlimited bet limits, and unlimited time, the chances are that we will eventually win more than we lose, even if we lose most of our bets.

Logic tells us that the pursuit of ideas that exclusively serve narrowly defined interests is a bad bet. Not that altruism is merely a clear-headed calculation, or that moving ideas out of labs toward humanitarian outcomes requires careful lessons in logic.

Before we met Thierry Marx, who makes his living imagining and serving up expensive meals, he was already doing culinary workshops with prisoners around Bordeaux; Fabrice Hyber, the contemporary artist, was planting and raising a vast forest around his childhood home to offer as a retreat to fellow artists; and James Nachtwey, one of the world's leading photojournalists, was risking his life to communicate through photography the human tragedy of war. Creators may not make much of it in the way they talk about creativity, but they often wish to see their ideas benefit others, and they spend their own time, resources, and creative talent trying to make it happen.

Why, then, do innovations so frequently benefit a few individuals rather than society collectively, or so easily turn into lifestyle conveniences rather than ways of preventing abject suffering, or serve those who wage war as opposed to the victims of war, famine, and preventable disease?

One good reason is the absence of a practical model for implementation. Innovations most readily meet the needs of those who are able to pay the price tag. In a free market economy, individual, corporate, and government interests combine to accelerate idea development and implementation when the commercial payoff is sensibly high or when some threat to large, well-represented interests is perceived—threats such as the security of nations or natural disasters. But there is little coordinated help for the person who invents a new vaccine capable of protecting the world's poorest against malaria infection. When the task is to scale up production of the new vaccine, distribute it to affected nations, and actually vaccinate those at risk of malaria, resources to implement altruistic change seem to dry up, if only because the probability of success, which assumes the concerted efforts of many interests other than the sufferers of disease, is so difficult to quantify. A credible collaborative model that directs the innovation to a practical and sustainable resolution of the humanitarian need often does not exist.

Where models of effective humanitarian change do exist, they tend to merge the interests and abilities of commercial markets, governments, and needy individuals in specially designed ways. A good example is the polio vaccine, first cultivated in human tissue by John Enders at Children's Hospital in Boston in 1948. This eventually led to the first effective polio vaccine in 1952 under the leadership of Jonas Salk, and, later in the 1950s, to mass U.S. inoculation, notably through a collaboration between the pharmaceutical industry and the March of Dimes. Toward the end of the twentieth century, the World Health Organization, UNICEF, the Rotary Foundation, and the pharmaceutical industry, along with other corporations, university labs, governments, and non-governmental organizations (NGOs), came together to tackle what the free market alone could not have done: making polio vaccination available to people in developing countries around the world.

Building hybrid, collaborative models that implement altruistic innovations with the efficiency and effectiveness of free-market models is clearly hard to do. Specialization in industry, academia, and government makes the task even harder. Rising to the challenge, many new NGOs have appeared in recent years, often guided by passionate leaders ready to pursue unconventional collaborative paths, aiming to address major humanitarian problems ranging from domestic violence to global warming.

In this chapter I show how artscience labs can effectively participate in the fostering of collaborative humanitarian models, by focusing on the experience through the 2000s of the nonprofit Medicine in Need, which applies advanced technology to address healthcare needs in developing countries. MEND is a small nonprofit based in Africa, Europe, and the United States whose roots started to grow in the idea translation lab at Harvard University. Guided by a combination of student enthusiasm and seasoned leadership, MEND's experience over the last decade illustrates how the goals of NGOs can complement the creative agenda of art and design experimentation, and how artscience labs potentially serve as a catalyst for humanitarian innovation.

⇒⇐

With the arrival of the new millennium, a wave of good will seemed to sweep through pharmaceutical science circles. There was a growing recognition that the most pressing health-care problems on the planet persisted in countries where the economic rationale of the pharmaceutical industry couldn't easily go. AIDS, malaria, and tuberculosis killed around six million people per year, mostly in places where people could not access the drugs and vaccines that routinely prevented these diseases in affluent societies. Many nonprofit organizations were organized to raise awareness and resources for combating these and other diseases, including the International AIDS Vaccine Initiative (IAVI), the Aeras Foundation for TB Vaccination, the Malaria Vaccine Initiative (MVI), the Global Alliance for Vaccines and Immunisation (GAVI), and the Global Alliance for TB Drug Development (TB Alliance). Alongside these major nonprofits, other NGOs emerged in a less collectively coordinated way, from One World Health (OWH) in San Francisco, founded by the charismatic Victoria Hale, to the recently formed Diagnostics for All, founded by the internationally renowned chemist George Whitesides. Medicine in Need grew up in this context.

I came to Harvard University in the early 2000s to teach, and ended up spending much of my early effort looking into how the aerosol technology I'd developed for inhaled insulin might be applied to the antibiotics used in the treatment of tuberculosis, a major global killer. A colleague at the University of North Carolina, Tony Hickey, had shown in some recent published research that inhaled antibiotics might be as effective against TB as injections in some cases, and might simplify treatment by, for instance, removing the needle from therapy for multi-drug-resistant tuberculosis. Motivated by the size of the need for diabetes health care and the associated insulin market, we had developed a new technology to make insulin inhalation therapy possible, cheap, and efficient; now the idea arose to apply this technology to a developing world disease where existing commercial markets did not drive research and development.

In my first class I asked students if my dream of combating tuberculosis through inhalation rather than injection of antibiotics truly made sense, whether it would ever find support within the international community. Several believed in the idea and began to look into what idea translation would require. Soon they came up with the notion to start the nonprofit, Medicine in Need. In retrospect, this initiative, and the deep motivation of undergraduate students, provided the rationale for what became the idea translation lab.

By the end of the semester the students were on fire to carry the idea forward and were arranging to travel to one of the most disease-burdened countries in the world, South Africa, where they planned to explore with local experts the mission and technology of MEND. Dan Yamins, the scientist of the group, was a brilliant student of mathematics who understood complex biomedical issues that other students generally took six additional years of schooling to master. Denise Kim, daughter of a former colleague of mine who had moved to Eli Lilly, was the engineer of the group. David Darst led the student group, which included Leif-Ann Reilly and Michelle Yu. Around twenty years of age, David already had demonstrated a brilliance for management, public speaking, and handling himself in an affably confident way that made others take notice. Somehow you knew that if David asked you to give to a cause, he was eventually going to give even more himself.

The student team traveled to South Africa, where they visited clinics and met with various scientists, including Bernard Fourie. Bernard later became the director of MEND in South

Africa. Through Bernard and other international leaders, the team eventually also convinced Gail Cassel, a distinguished scientist at Eli Lilly, to support the MEND mission. Gail later joined the MEND Board of Directors.

Over the next few years MEND grew by scouring for resources wherever it could find them. With the leadership of the MIT-trained scientist Jenny Hrkach, the organization received research dollars through a National Institutes of Health grant to my Harvard research group. Two years later it began to receive direct and indirect grants from the Bill & Melinda Gates Foundation. Bernard Fourie opened the MEND office in Pretoria, South Africa, in 2005. By then, the original student founders had gone on to pursue other careers in medicine, mathematics, and venture capital. But new Harvard students kept the passion alive. In every class I later taught, undergraduate students wished to learn about how their education could help them participate in addressing humanitarian problems anchored in the plight of the world's poorest. Obvious business models for most of the ways we came up with didn't exist. But this didn't stop student ideas from pouring into the idea funnel.

MEND continued to grow. Two years into the Gates Foundation grants, the organization recruited Andrew Schiermeier as CEO. At just forty years of age, Andrew was a Harvard-educated applied mathematician turned pharmaceutical consultant, venture capitalist, biotech CEO, and now entrepreneur at the helm of a young humanitarian organization. When he took over the leadership of MEND, the same year we opened Le Laboratoire, he cut his handsome CEO salary by a third. Our board, largely made up of business-savvy pharmaceutical leaders, had difficulty understanding why he made the move: CEOs of nonprofits like MEND typically joined after long careers in the pharmaceutical industry, not in the middle of those careers. Moreover, MEND had about a year of funding, with no guarantee of more to come. How would it pay the bills?

For about a year after Andrew arrived I tried to argue—through professional conferences, art exhibitions, and articles—that surprising, impossible-to-foresee approaches to global health financing and delivery might come by inviting artists to engage in the problem creatively, not simply as vectors of communication. But months after Andrew joined, in the fall of 2008, my efforts had not resonated with anyone else as far as I could tell. The fundraising role of celebrity artists like singer Bono of U2 was clear enough. But the argument that art and design would help us in a different, more fundamental way did not catch on: it seemed to distract potential donors toward "process" at just the moment when attention needed to be riveted on product.

Andrew's wife, Marie-Cécile, had grown up in the South of France, and one consequence of this was that Andrew traveled from Boston to Paris every few months. Each time he visited Le Laboratoire, he walked around the exhibition spaces and spent time with the team—in the office, at the café, wherever we all happened to be. He agreed with the premise that creative process mattered to the development and delivery of better health care for underserved populations, but he was too realistic not to understand that we would need concrete results before anyone invested.

So he went to work on other development strategies: building new bridges with the pharmaceutical industry in developed and developing countries, deepening the relationship between MEND and Harvard University through the new Wyss Institute for Biologically Inspired Engineering, and keeping an ear attuned to students who came through Harvard's idea translation lab.

The economic crash in the autumn of 2008 led to more belt-tightening within the humanitarian pharmaceutical sector, inside and outside the Gates Foundation, which was by then the sector's most visible source of funding. This decline in fortunes accelerated worry over our still-uncertain business model for developing drugs, vaccines, and delivery technologies for diseases of poverty.

But students in the idea translation lab didn't lose sleep over these worries. Encouraged to think broadly in the arts and design, they started to come up with ideas that eventually might lead to sales of original products in the developed world with distribution of useful products in the developing world to create a sustainable business model for improving global health. Three of these models stood out. One related to vertical farming, another to soccer, and a third to water. Andrew learned of their ideas at our innovation workshop in Paris in the summer of 2009 and began to develop a new strategy to link MEND with the network of artscience labs in Cambridge, Paris, and Cape Town.

≫≪

At the turn of the millennium, around 36 percent of Africa's one billion people had no direct access to clean water, and 75 percent had no clean water piped into their homes. Annual deaths in Africa from diarrhea-related diseases, aggravated by contaminated water, were estimated at 707,000 in 2002. By 2009 diarrhea or dysentery accounted for one in five childhood deaths worldwide. To access available water in the most water-constrained regions, people walk several miles each day to carry water in jugs, old gasoline containers, reused water bottles, and various other vessels. The size, cleanliness, inconvenience, and stigma of head-borne water containers combined to make water transport the task of women and of children above the age of ten who lived with their mothers.

In the fall of 2008 students in the idea translation lab tried to come up with a novel way to solve these problems and improve health in low-income water-constrained regions of the world. It started with the idea of "tensegrity"—the architectural principle whereby structural elements join by tension rather than pressure. A balloon is a simple example of a tensegrity structure. Don Ingber at Harvard Medical School has long advanced a theory that the cytoskeleton of biological cells is also a tensegrity structure (as I described in *Artscience*). With the Wyss Institute supporting student project development, I proposed to work the concept of tensegrity into my class. Might the biological structure of a cell, particularly a tensegrity structure, be relevant to improving global health? Michael Silvestri and his classmates, with the encouragement of Hugo Van Vuuren, started to look at how this idea related to water transport in Africa, a problem Hugo understood well from his upbringing in South Africa.

The biological cell is a fascinating water transporter. Water enters and leaves the cell by passing through the cell membrane, and as it does it gets filtered. Salts and other substances that can be toxic to the cell are kept out in the natural process of water transport across the membrane. Cells also move through the body in a variety of ways, piggy-backing on the transport of blood through the circulatory system, rolling through tissues by a phenomenon known as chemotaxis, a kind of chemical tether that pulls cells along, and by other specialized means. Cells swell and shrink, flatten and elongate, changing their form to change their function. What a fantastic notion that we might use these principles to engineer better ways to carry water!

Just before mid-semester presentations, I met with the students and we decided that the idea had progressed to the point that it merited a patent. We headed over to the Border Cafe in Harvard Square, where we wrote up the patent disclosure over chips and salsa. The next day we submitted the disclosure, and soon Harvard University filed a first provisional patent. The semester ended with much enthusiasm. But it remained unclear precisely how to make these kinds of water vessels in a simple and practical way that could be useful in water-constrained parts of the world. Michael and his classmates had made structures with wooden rods that behaved in the way the tensegrity structures first imagined by Buckminster Fuller did, but it seemed impossible to manufacture these complex structures cheaply and on a large scale.

I asked Michael if he wanted me to involve a designer. He did, so I raised the idea with Mathieu Lehanneur. What do you know of tensegrity? I asked. It seemed a complex matter, as far as he could tell. For a few weeks we played around with structural ideas. Then one day in the late spring of 2009 Mathieu drove by my office on his scooter. He sat down and we talked. Classical tensegrity structures, with solid rods that connected in ways that allowed you to collapse the structure and have it spring back to its original form, seemed too complicated to manufacture on a large scale, and too precious to imagine using in rugged settings.

Mathieu walked up to the white board in my office and sketched an alternative structure. The new structure—we called it the Pumpkin—had principles in common with the original idea. As with the cell, form related to function. You squeeze a cell and water comes out, or you flatten it and it grows differently; the Pumpkin did this, too. Water enters and leaves through the cell membrane and is filtered as it moves through; the Pumpkin did this also. Like the cell, it had multiple ways to travel, depending on what was available to carry it. The Pumpkin was simple and elegant as well.

The Pumpkin had the collapsible nature of a Chinese lantern. When you unfolded it completely, it was spherical. I suggested we might place a cylindrical water filter on the axis of rotation, which Mathieu drew on the board, creating dimples on the top and the bottom. In its pumpkin form the water object sat on your head, or could be rolled on the ground, like a wheel. Half open, it could cling to your back. We came up with a simple way you could open it up and clean the inside. The intriguing thing about this object was that it could also hang like a purse from your shoulder. And that suddenly looked like something anyone might use.

Mathieu's studio made a paper prototype of the object. In June 2009 Hugo traveled through Tanzania and South Africa, working on his Lebone project and joined by Julien Benayoun, a young designer who worked with Mathieu. In their travels to rural villages, Hugo and Julien discovered that villagers liked the Pumpkin. They appreciated the novel options it gave to water transport—on the shoulder or on the back—and the way it looked remarkably contemporary when sitting on your head. It seemed fashionable. A few women interviewed in an affluent part of Johannesburg liked the Pumpkin, too. They could imagine carrying water around in it when they left their homes.

Later that summer at the innovation workshop in Paris, we sat down with Michael, his classmates Mathieu and Julien, and Andrew Schiermeier. Noemie Tassel, who headed up MEND in Paris, joined us too. Progress seemed encouraging. But water was fast becoming the major problem of the planet. Many thought about solutions. Would this one really add value?

That fall, with the encouragement of Andrew, I visited Pretoria and sat down with Bernard Fourie and Philip De Vaal, an engineer in the Department of Chemical Engineering at the University of Pretoria. The Pumpkin was just an idea, mostly a hypothesis. You couldn't say it would improve health in water-constrained regions any more than you could say an exhibition by James Nachtwey would galvanize efforts to fight the war against infectious disease.

The surest value of this idea lay in the process of its translation. Perhaps the Pumpkin would be a catalyst for local village innovation, wherever that innovation took you. This line of thinking led to the startup of a new artscience lab in South Africa, supported by Harvard University funding.

The Harvard Institute for Global Health (HIGH) is a group of faculty, researchers, and staff across the university dedicated to improving global health through education and research. HIGH was created with the vision of the renowned tuberculosis researcher and professor Barry Bloom in the early 2000s to aggregate the diverse research and practice that took place at the Medical School, the School of Public Health, the Business School, the Law School, the Kennedy School of Government, and Harvard College around global health crises, ranging from infectious diseases to potable water. I had been on the Faculty Executive Steering Committee since 2004 and active in the debate around mission, which certain board members wished to orient toward the creation of knowledge and others toward improved health care. A few, including MacArthur Award winner and public health researcher Sue Goldie (who later became HIGH's faculty director), fought for a compromise. This group held that, since the core mission of the university was neither the creation nor the application of new knowledge but, rather, education, the mission of HIGH might be better defined as educational—specifically, a kind of experiential education where students would learn through developing original global health ideas. As we gained evidence through the idea translation lab of students' passion for global health research and development, the Executive Steering Committee decided to give the idea translation lab a role in HIGH's educational agenda.

This happened around the time the Pumpkin idea started to gather momentum. Amanda Brewster of HIGH and Hugo Van Vuuren wrote a proposal in the fall of 2009 to create a branch of Harvard's idea translation lab in Cape Town, South

Africa. The organization would receive students each year who came to translate ideas related to global health, often working with African students and innovators. The first pilot project was the Pumpkin. Funding arrived at the end of 2009. We had the final partner we needed to start a three-year African artscience experiment around water transport

Noemie Tassel assumed responsibility for global management, while Bernard Fourie and Philip De Vaal assumed responsibility for student engagement at the University of Pretoria and management of South African village work. Hugo and Jessica Lin, representing the idea translation lab at Harvard and Cape Town, worked with Michael Silvestri to oversee other village work around Cape Town. And I worked with Mathieu Lehanneur, Julien Benayoun, and Manuella Passard of the LaboGroup to develop a working prototype of the Pumpkin.

The goal of the experiment was to test whether an inexpensive, practical, innovative design with multiple modes of transport might lead to greater frequency of human traffic between clean water sources and human populations, and therefore greater volumes of clean water transported. As important as this outcome was, however, even more critical was the process we hoped would lead to it. We would begin with the Pumpkin, but, working with villagers, we hoped to help local youth and entrepreneurs think creatively about their own water challenges and work toward local solutions, possibly related to the Pumpkin, possibly not. Meanwhile, we would approach retailers in Europe, the Americas, and Asia who might have an interest in selling the Pumpkin for water transport in the developed world, so that we might generate revenues to support the charitable mission. With a bit of luck, the three-year experiment might be able to finance itself.

By late fall 2009 I came up with a related idea, which I shared with the French designer François Azambourg. The biological cell has a further interesting property. Its container, the cell membrane, cannot be dissociated from what it contains—the cell fluid and its cytoskeletal microstructure. It is the same with all biological life: you cannot simply open up the life form, pour the life out, and fill it up again. But that's precisely how we make bottles and glasses and other fluid containers. The container is ultimately autonomous from the contents.

What if we changed that? It might lead to a more natural, ecologically rational approach to carrying the liquids we drink. I asked François to consider what this might mean to the future of bottle design, and we decided to do a new Laboratoire experiment. We aimed to exhibit whatever results we generated in the fall of 2010, which would give us time to move the Pumpkin idea forward. Ideally, the two experiments would together drive attention to the charitable cause.

Time was short. We looked into ways we could make bottles that looked like different fruits, were composed of juices, and could be popped into your mouth, so that you would "drink by eating." That became the slogan. This idea seemed to us like a new platform for research and development in the FoodLab, so we invited the FoodLab scientist Raphael Haumont to become involved, and we began to talk about it all with chef Thierry Marx. My postdoctoral researcher Sidi Bencherif also threw his hat into the ring, as did Don Ingber of the Wyss Institute. This was the tenth experiment of Le Laboratoire.

≡≡

We're still in the early stages of design development for the Pumpkin and related water ideas. If all goes well, we hope to be soon selling Pumpkins in French stores and bringing Pumpkins to rural villages in Africa, learning from cultural incubation. Will it translate into a healthcare improvement? Will sales in the developed world pay the price tag for humanitarian efforts in the developing world? Perhaps the idea will produce a sustainable approach to water transport that we cannot right now envision, one imagined by innovators whom we have not even met. We can't know, which is, of course, part of its fun.

As I write, Manuella Passard is preparing with Chinese and French manufacturers a couple of prototype Pumpkins for testing in South Africa in the coming weeks. We've changed their design at least ten times since Mathieu and I took a weekend boat trip together eight months ago. It does not look much like the object the students initially envisaged, and the Pumpkin, as it will be produced this summer, is surely just a first step toward what is needed. So why has this idea, more than any other we have developed in artscience labs over the last few years, generated such collaborative interest across expertise areas and organizational missions?

Experimental altruistic ideas defy the kinds of perceived conflicts of interest that sometimes stop idea translation in non-lab environments with more narrowly defined interests. The general importance of the dream trumps the logic of organizational mission, so that we are freed up to forget for a while our specialized organizational objectives; working in a lab, we are encouraged to experiment, given the right resources to bring those experiments to a place where others can react, and able to receive feedback from others, who recognize that our immediate goal is experimentation itself.

Structuring our institutions around narrow missions of learning, profitability, or aesthetics, we sometimes lessen the connection between institutional mission and the collective needs these missions aim to address. Artscience labs are one way to help strengthen that connection—by lowering institutional barriers that prevent idea translation from going viral.

7

CREATIVE BANDS

We create best, and longest, when working with others who challenge, encourage, and generally help us better articulate and develop ideas. Dreaming with others, creators are less conscious of risk and freer to pursue avenues of inquiry that lead to surprising outcomes.

Creative bands are about as integral to the innovation model of artscience labs as they are to the design company Ideo, or to media organizations such as Google, the MIT Media Lab, and Ars Electronica Futurelab. All of these innovation organizations count on creative bands to accelerate the movement of ideas from conception to realization. In artscience labs, most particularly, creative bands tend to gather around a single creator, as they do in an art studio or a science lab; they aim less at client desires than at perceived needs or opportunities, which can change; they frequently have little experience or training; and the dreams they pursue can be diverse, with outcomes ranging from a work of art to a work of social engagement.

Creative bands, wherever they appear, are usually small enough so that each member of the band can preserve and express an individual identity, yet sufficiently large so that the aims of the band can still be achieved. What is the ideal size? That depends on the nature of the band. It is hard, for example, to imagine The Beatles, one of the most famous creative bands of all time, producing innovative music as a band of thirty or forty. When we think of The Beatles, we think of the individuals—John, Paul, George, and Ringo. We needed all four. The Beatles would not have existed with one fewer. Yet being small in number, each of the Beatles was recognizable, knew that whenever he wrote a song, played a new riff, or delivered a funny line on stage or in a movie, the result would be something "the Beatles did."

If creative bands remain limited in size, they do sometimes multiply in number. The Bauhaus did not grow to a very large faculty or student size over its fourteen years of existence. But when the Bauhaus ended, its leaders went on to start other art and design education programs, which expanded the international reach and impact of the Bauhaus model. Walter Gropius became chairman of Harvard's architecture school, Josef Albers went to the new Black Mountain College in North Carolina, László Moholy-Nagy ended up at the New Bauhaus in Chicago, and Mies van der Rohe became head of the architecture program at the Illinois Institute of Technology. Startup companies coming out of Stanford and MIT in the 1970s, to cite another example of a creative band, inspired investors, creators, and universities alike to produce other spinoffs from successful academic labs that eventually aggregated in Silicon Valley and along Route 128 outside Boston.

Creative bands have this magical ability to reproduce virally, as if human civilization had been waiting to translate its creative energies into a better world and just needed to learn how it was done.

The other day I attended an end-of-year event at my sons' elementary school in Paris. A Saturday morning, sunny, cool breeze, my wife, Aurélie, and I trying to track down our three boys on the urban campus, a ten-minute walk from the Luxembourg Gardens.

Jérôme runs up to me first. He is ten, tall, athletic, independent, first child. Throws his arms around me and wants to introduce me to his friends. They all stand in a line ready to head off to their ceremony, where the school director is about to distribute awards—best artist, best behaved, best grades. Jérôme has a healthy conservative slant on risk-taking. He takes risks all the time, of course—you probably can't avoid it at ten years of age. Some of these are imposed (a change of schools, cities, countries three times in the last three years); others are elected (he started playing ping-pong at the beginning of the year, at a few tables on the playground, and climbed to the top of his class, patiently moving from early mistakes to the skill level he shows today).

Jérôme's friends, boys his age, look my way, guardedly. His good friend Antoine shakes my hand. Monsieur, he says. The others watch. Jérôme's band. Similar in size and smartness, they gather together each recess, no single leader, though each is capable of it. Their collective ambition seems obvious: they want to have fun, as most anyone their age does, sports being a primary outlet, and they want to do well in school, finish in the top five or ten each quarter. They're old enough to see the danger of falling behind in school, having your name read out in class, low score, you could have done better, and everyone looks your way, registering this fact about you, which will take days, weeks, to erase, and maybe you never will. There's a punishment to French schooling if you're not careful about it—anything you do, or don't do, is liable to be paraded at the end of the semester before your classmates. For Jérôme's band, the goal of school seems to be to survive it, having as much fun as they can along the way, but they aren't necessarily worried about figuring out what it all means.

"Figuring it out" is the goal of the band that has formed around my youngest son, Thierry. Thierry is six, youngest kid in his class, no longer the smallest but not far from it. Thierry and his friends are still making up their minds about how to behave. They have a band that forms around a clear leader, a boy named Antonio, who is tall, strong, occasionally gets into skirmishes, but does well enough in class to seal his leadership status. For Thierry, Antonio provides cover. He sometimes says the provocative things Thierry would like to say, just to see how his teachers and friends will react and to learn from it. That behavior is part of the core mission of Thierry's band—to figure things out by a little provocation short of disaster.

My middle son, Raphaël, is eight, and running around a tree, free-spirited, momentarily confident, good-natured, and consistent. Raphaël's band, a group that includes Thibault, Aliénor, and Paul-Henri, is leaderless. Their teacher wants to keep them together next year because they're good for the class, high achieving and gentle, a refreshingly curious band, with a mission neatly positioned between the bands that have formed around Raphaël's older and younger brothers.

Nothing in my sons' particular schooling imposes this kind of banding. There is even something rebellious about it. In the classroom our kids are told where to sit, how long to sit, what to write, where to write it, what to say, and in many ways how to "be" with a kind of extraordinary precision we would never think of imposing on a non-incarcerated adult. Recess becomes an oasis. Later the oasis will be afterschool hours, summer vacations—a refreshing time and turf where social, educational, and professional constraints disappear. The kids group together, or not, with their friends, in a way that is instinctive, that comes and goes. Self-assembly into bands is a fundamental process by which they grow up to understand their world and their place in it, while they acquire a formal education on the side, the two objectives precariously aligned.

The forces that group Jérôme, Raphaël, and Thierry along with their friends will grow less obvious over time. Yes, they will all continue to wish to play, to understand, to survive grueling periods. But other forces associated with specialization —the need to progress along specialized pathways of learning, finally accepting, on faith, that these paths will lead to the joy, understanding, and survival they seek—will likely become overwhelming. Eventually it will become difficult to meet up with just anyone at all to "play," or to "explore," or to "hang out."

Specialization is one of civilization's ways of orienting individual brains for the benefit of society as a whole. We cannot, for instance, do without civil engineers who see clearly how to build and maintain the infrastructure of our cities and country. Obviously, they need to acquire increasingly greater knowledge about various forms of energy generation and consumption, new materials of construction, the latest data transmission systems. We don't ask our civil engineering students whether they're interested in studying these things; we determine, collectively, that to become civil engineers students have certain knowledge and experience needs. These needs are myriad and the list grows every day as technology advances and new challenges arise. So we create subdisciplines within the field of civil engineering that allow students to learn even more specific knowledge sets. And if a civil engineering student takes courses in the field of energy, say, within an accredited program, we assume that he or she will find a good-paying job and wind up positively contributing to society's civil engineering needs as they relate to energy. Specializing in energy will, of course, cause our student to associate with other civil engineers studying energy, and, later on, with employees of businesses that need civil engineers with a specialization in energy. This is how specialization guides us as we grow up.

It's not that there's anything particularly forbidden when energy specialists (to keep with the example) hang out with, say, visual artists. It's just that hanging out with visual artists is not a collective priority we have set for civil engineers (why should we?) in comparison to priorities we have set for them to study energy, land jobs, and put in long hours at the office.

Jérôme may become a teacher of history, or, who knows, a ping-pong champion. But most likely he'll have to choose one over the other. And his friends, schooling, reading material, and work environment will all probably reflect his professional choice. This decision to specialize, which entails a decision to seek out instructional environments that turn some children into history teachers and others into ping-pong champions, is to risk choosing a way of life that will maintain the constraints of our childhood schooling all the way to the grave. The specialization of education, industry, culture, and society is understandable and too often necessary, but to the degree that it becomes the dominant educational force acting around us, it implies that we know more than we actually do about how to divide up human knowledge, activities, and aspirations, and how this will matter to us, individually and collectively, now and forever.

Creative bands are mostly, though not always, a youthful thing. Those who usefully buck the specialization system—those we think of as innovators—eventually figure out how to reconstitute such bands long after the rest of us have fallen into the trenches of our specialized lives. Within the collaborative structure of creative bands, they explore, innovate, and achieve meaningful ends that they could not achieve if they worked alone. These may be music groups, startup enterprises, theater companies, any organization that creates things a society doesn't count on but will eventually recognize a need for. It seems fair to assume that no form of specialization can exist to guide young people in this direction. And fortunately, because these things that will be created by innovators are needed, our specialized societies are not so lost in specialization that they have precluded this banding from occurring. Space exists for little creative groupings—and here is where labs come in.

≡≡

It is easy to interpret the lab as the environment that helps some leader to assemble a group for the purpose of developing an idea, and who somehow knows, in good times and bad, how to impose a compelling vision on everyone who joins the band, remains in the band, and needs to evolve with the band's idea. The lab is there for the chief experimentalist, and everyone else appears to lend a technician's support. In this view, banding, and the focus of the lab, can be reduced to the persuasive power of a charismatic leader. However, if collaborative idea translation were this simple, band members would be more interchangeable than they are. And successful labs would be easier to find. In truth, the most ambitious ideas, the greatest innovations, require that each member of a creative band possess the idea in a deeply personal way, one that causes each member to be watchful, to react autonomously, indeed at times to act as the leader of the band.

Vision derives from some hypothetical idea, the dream, espoused by everyone in the band, an idea that requires belief, though all the sufficient supporting facts are lacking. This vision has mystical qualities. Leaders of creative bands tend to imagine as realizable some collectively meaningful dream and to analyze or to plot out concrete paths to realization, seeing how to motivate everyone else in the band to do something definite while maintaining the passion and enthusiasm associated with the dream.

And the fused process of dreaming and analysis that reigns in the minds of creative bands—what I call artscience—is, while natural enough in the young, so foreign to the way we adults think in our specialized world that we tend to see it as magical, otherworldly, what Robert Pirsig calls Zen, something near to what Mihaly Czikszentmihalyi calls flow. To be visionary is to believe completely in something that many, with good reason, do not believe in at all.

To be effective, a creative band must share some long-term, meaningful dream. It might be to create a new theater to respond in some artistically sensible way to certain obvious changes in social discourse, or to commercialize some new product that will make life, in however small a way, more enjoyable. While the dream is never forgotten, generally it does not dominate daily conversation, remains mistily removed from immediate concerns, and so, while the band's mindset forms around a long-term shared dream, it aims at short-term goals, which, everyone assumes or has been told, will become indispensable to the achievement of the dream.

Creative bands—however enjoyable it may be to participate within them—don't form in the first place around individual interests. Even personal investment groups, when creative and successful, take pride in their collective performance. Those who beneficially join the band love the idea of creating something that will matter on a scale larger than their own social or professional circle. And good leaders make that collective interest apparent even in the detail of individual, everyday tasks, often by exaggerating the nearness of the shared dream or the collective importance of the individual tasks, or both. But leaders are not as singular as we believe, for we all need to become leaders—rebels turning our backs on rules that exist for each of us individually—if we wish to reclaim the freedom of the creative band.

I came up with the idea for the product we call Le Whif in the summer of 2007, a month before the fall semester began at Harvard University. But clearly I could not have realized the idea of "breathing food" without help from others. First, I was very busy. I was teaching, writing, raising a family, and opening Le Laboratoire. But even if I had nothing else to do with my time, I still could not have realized this idea alone. I knew about the physics and engineering of aerosols and could imagine working out various technical hurdles on my own, but I didn't know much at all about flavor or haute cuisine, and even less about food manufacturing and marketing—I was ignorant about a thousand things that would be critical to the success of Le Whif.

I didn't even know for sure if this was a good idea. So it occurred to me to ask my students. To be fair, I didn't simply ask them; I put the question to them in the form of a hypothetical idea—a dream, in fact, which seemed to me likely to excite them. I told them I was ready to share my vision and its development, so that, if they wanted, they could learn what it means to translate an idea and maybe go on to participate in an exhibition, even start a company.

When I presented the idea to my students in September 2007, I deliberately inflated my "vision." One day we, humanity, might eat by breathing, I said. Eating is a basic human thing. Breathing is basic. Nobody has ever imagined that eating and breathing could merge into the same act. This idea, simple and broad, intrigued a handful of students, who formed a creative band. For a few months this group excitedly gathered outside of class and tried whiffing carrot powder, among other things. Nobody told them exactly to do most of what they did, even if I met with them regularly and gave them advice. More than anything, they worked out of passion for the idea, which, as the band's leader, I tried to keep alive throughout the temporary discouragements (whiffing pepper was a very bad turn), covering the seemingly unrealistic nature of the idea (coughing was one of our biggest problems) with whatever credibility I retained as their professor.

It helped that I presented this idea as part of a regular class, however unusual that class turned out to be. The students were truly interested in breathing chocolate, but they almost certainly wouldn't have pursued the idea had they not had the promise of receiving a grade at the end of the semester. This brings up a second characteristic of creative banding, beyond its formation around a common dream: we use organizational infrastructure—we even create infrastructure (as I created my class)—to facilitate the formation of creative bands. The particular bands that formed around my sons at their school would not have happened if the kids did not attend the same school, if the school had not allowed long recreation periods, if there had not been ping-pong tables and lots of grass for play, and so on. I wouldn't argue that recess at my sons' school was the principal catalyst of band formation, but leaders do take advantage of resources at hand, just as I took advantage of my class to create the first creative band around Le Whif.

With the semester drawing to an end, whiffing still seemed like a reasonably good idea, and as a result the circle of our band widened. This was inevitable. Nobody in the little Harvard band knew very clearly what to do next with Le Whif, or would have had time for it even had they known. They were, after all, undergraduate students, had classes to attend, papers to write, exams to take. They'd also never manufactured plastic inhaler parts or probably even heard of the notion of failure-mode testing. Beyond this, they had no financial resources to invest in the idea and no realistic sense of how to bring Le Whif before the public palate in a responsible way.

My friends at LaboGroup in Paris entered the band. Our mindset changed when they did. The collective long-term dream remained this idea of beneficially breathing food, chocolate in particular. But the immediate objective, comparable to my students' objective of delivering a final class presentation, narrowed down to a spring exhibition. We were planning to open an exhibition of Thierry Marx's work in April 2008, which involved his collaboration with the French colloidal scientist Jérôme Bibette. That exhibition needed significantly more material than it currently had, all of us at Le Laboratoire felt. Talking to Caroline Naphegyi, we came up with the notion of a Whif Bar. José Sanchez managed to interest Nespresso as a sponsor. So the message that motivated the band now became to create inhalers of chocolate to display and use in a culinary art exhibition at Le Laboratoire in early April.

This vision excited at least two Harvard students, Larissa Zhou and Trevor Martin, who flew to Paris in April to participate in the trial of an idea that had emerged just a few months before in their class. For our team in Paris, enthusiasm for Le Whif was high because it meant, among other things, that we would have something extra to show for our third exhibition. Maybe something commercial would come of it, but, truthfully, we didn't have time to think seriously about what would follow the exhibition.

Here is a third thing to note about creative bands. If they are going to be successful at moving an idea ahead to some sensible degree, creative bands will need to evolve in structural, compositional, and functional ways, simply because the idea itself will grow and evolve over time. A band's success at navigating this evolution usually requires someone, and ideally many people, who understand perfectly the culture that previously existed, and understand—at least well enough—the culture that is setting in, to reassure those who will evolve with the band's vision that the idea of the group will not be lost, that, on the contrary, it will actually be saved.

This transformation becomes one of the greatest challenges to realizing ideas through creative bands. New members wish to own the idea of the band as much as anyone else owns it, and this means that existing members must be willing to see their idea change with the new members' perspectives. And, probably without much experiential basis for trust, the old members must believe enough in the motives and reliability of new members that they will listen to and share with them almost as if they were brothers and sisters of the same family. The arrival of new members means that the band was previously incomplete, or suddenly became incomplete, and at a minimum it implies that the old members were inadequate to the mission of the dream they espoused. The way of communicating between members of the band changes, and maybe the context of communication changes too—a new corner of the playground, or a new city.

In our case the growth of the whiffing band to include José and Caroline and eventually many others would not have happened as seamlessly as it did without Jonathan Kamler, who had studied at Harvard. He was working for LaboGroup while living in Cambridge, and, having French family on his mother's side, would be traveling to work in Paris at the end of the school year. Jonathan knew perfectly well the culture of the Harvard student band, since he'd been central to its creation, and he also knew the culture of the Paris band, or knew it as well as you could at his age and without any directly relevant practical experience. Among other very constructive things that Jonathan did in the winter of 2008, he built confidence and dialog between the old (student) and new (LaboGroup) bands.

José made design recommendations to facilitate manufacturing. He figured out the kind of chocolate that could be whiffed and found sources for the flavorings. Caroline came up with the idea of the Whif Bar and orchestrated the exhibition in the spring of 2008. Jonathan had coached the students through aerosol testing in my lab and had helped them make and test their first Whifs. Now he helped them organize their trips to Paris, where they would work with him to fill the newly manufactured Whifs, which happened to be very fragile, with chocolate powder.

It helped that all this happened very fast. In successful creative bands—to enumerate a fourth characteristic—anticipated things happen speedily. The will of the band aligns and is emboldened in the process of action while it often splinters in the lull of a long wait. A public exhibit! It was just the thing to build consensus!

This encouragement was not very directional. Directionality kills the millions of instantaneous ideas that need to arise in the minds of a creative band as it works speedily toward a goal. Working for a leader who imposes tasks makes you attentive to instructions anchored in the past. In an environment that evolves as rapidly as a creative lab must, what you're told to do is almost never exactly what you should do, given the reality of the present, which you can never precisely know a priori. Like sailing a boat through a storm, pretending to know what to do ahead of time invites disaster. Micromanaging an artscience lab pushes the creative band away from the passion inherent in the idea they are helping translate and breaks down communication with creators. This can be necessary in the short run, as it was in the days just preceding the exhibition, but it moves the act of creation toward that of production and the artscience lab toward the role of factory.

Whiffing in public was a tremendous risk. We'd never whiffed with the manufactured inhalers, and they arrived only a few days before the exhibition. We had essentially done no product testing, had no idea if people would love, hate, or be indifferent to our idea. Making risky bets in the course of idea translation is the fifth principle that makes creative bands function successfully. Smart risk-taking unifies a group. This is almost always true when the bet pays off, and with good leadership it should be true even when the bet is lost.

Things went smoothly on the night of the exhibition opening, but after that, the band saw its hopes dashed many times. We manufactured Le Whif with defective molds, missed manufacturing deadlines, failed to explain clearly to the public what we meant by Le Whif, or who would ever use it—we did a hundred things wrong. And each time, we of course learned something. As leader of this group, I did my best to articulate what that learned thing was, which I presumed to know even when my opinion changed by the day, and to make sure everyone understood why we were actually better off than before, and still on the hoped-for track, still pursuing the dream.

The day before we commercialized Le Whif, on April 29, 2009, Jonathan emailed me on my way to the airport for my Paris flight to say that the product didn't work. The problem lay in the way Le Whif was supposed to open. It either failed completely, or didn't open to the necessary degree. I called José. Go try it yourself, I said, looking despondently out the window at cars rushing past. Open all the packages, do whatever you need to do to get the maximum number to work, and throw away the rest. José said he'd thought about this, too, and it was reassuring that we saw things the same way; sharing intuition made us more decisive to act on it than we might have been alone.

Failure, in this sense, happens constantly in any ambitious idea translation, and it helps to spot it early, to actually look for failure even when nobody is pointing it out. Because aside from avoiding the idea-killing disasters that follow a series of failures (and failures often happen in sets, sending us a forewarning if we're paying attention), failure fortifies a creative band, building the tension it needs to be reactive and avert sudden catastrophe.

Sometimes, when it can't afford to fail, a band will go to ludicrous ends to avoid that outcome. This was the case on the day of our April launch. The pressure of the moment—and, truly, these first Whifs were bad—led Jonathan to what anyone would have to admit was a rational conclusion. With all the media attention we were receiving, with Internet orders piling up, he concluded that we had to take Le Whif off the market. We couldn't sell a defective product to customers or hand it out to the press corps. But the conclusion we needed to draw was not this logical one, rather it was an irrational one: yes, we would sell, we would find a way to help people whiff with pleasure, and we would avoid the disaster of a product that malfunctioned.

In the end, all worked out, and we launched the next day with a successful press conference, a lunch with dignitaries from around the city, then a party in the evening. Nobody seemed to notice that anything was wrong, or, if they did notice that the Whifs were far less than perfect, the momentum of the idea translation mixed with a little champagne helped them not pay much attention to it.

All this crisis detail may obscure the fact that lab life is, for those who enjoy experimentation, actually terrific fun. In fact, it is far more fun for creative bands to be in crisis than not, particularly when it is a kind of self-imposed crisis that the band can collectively believe will pass because the cause is so beautiful, the team so aligned in their commitment. Creative bands then rally around ideas as they advance, sometimes better than they hoped, and sometimes less well. Management steers the boat actively, no autopilot, and leadership is never far from the wheel.

≡≡

We navigate the transitional moments of the idea translation process by inductive association and deductive reasoning, comfortable with uncertainty and yet able to identify large problems and reduce them to meaningful small problems that can be solved. Art as process, and science as process, fuse as one—the core process of creativity. Creative banding helps us navigate these risky periods of innovation through what I have just illustrated to be five critical stages of the experimental process:

◆ Someone brings the group together around an idea translation vision.

◆ The leader looks for and uses pre-existing organizational infrastructure to help the band coalesce, evolve, and grow.

◆ The creative band grows larger and its culture evolves as the idea advances; those who understand the culture that previously existed and the one that is setting in reassure members that the idea of the group, its vision, will not be lost, that it will actually be saved.

◆ The idea develops quickly, and this helps align the will of the band, which may otherwise splinter.

◆ The group takes risks to translate the idea, and risk-taking unifies the creative band.

The lab's leaders direct the creative band, helping it evolve as the idea moves along, even guiding the band to understand that unless it also changes, the idea will die. A will to guide ideas beneficially to their next destination is, to the lab, what the will to make a profit is to a free-market enterprise. And it is intuitive only to those who understand and share the philosophy of experimentation.

Creative banding to develop innovative ideas happens best under particular conditions, which, if not present, make it difficult for even a Bill Gates or a John Lennon to carry forward ideas and realize innovative dreams. Once these conditions are present, however, the human passion to create and effect beneficial change often leads to viral growth of the model. This was the case with Silicon Valley during the dotcom era, with the Motown label during the early days of R&B, with Broadway in its heyday between 1945 and 1965, and with the American translational science lab during the half-century after World War II.

Viral growth of innovation in the arts and sciences does not happen just because a particular creative enterprise grows in size. A dotcom startup, a music group, a theater company, or a science lab all have limited capacity to create, and pushing that capacity too far can diminish its originality and eventually eliminate the conditions required for innovation. In the network of labs described in this book we have, today, twenty-four experiments under way in Boston with urban teens, another eleven experiments under way in Cambridge with Harvard students, six experiments continuing from last year, two experiments reaching fruition as exhibitions at Le Laboratoire this fall, three more experiments planned for exhibition in Paris in the winter, spring, and fall of next year, and two experiments that are early-stage commercial products. That makes forty-eight experiments at different stages of conception, translation, and realization, all with the inherent uncertainty that experimentation entails. This network probably cannot serve many more ideas than this without losing the small, nimble, youthful spirit critical to the creative band. Other models might do better. Those who worked in the studio of Pablo Picasso, for example, generated, translated, and realized over 20,000 works of art in the course of his career—an almost unimaginable body of work.

Viral growth of innovation happens mostly by replication of conditions that allow creators, working in bands that form in many laboratory contexts, including art studios, research labs, and corporations, to translate innovative ideas. Artscience labs represent a new form of experimental environment focused on art and design innovation. Their appearance in recent years in various cities around the world seems to reflect several conditions favorable to creative banding, including those I summarize below:

Artscience labs expand the possibilities of laboratory experimentation beyond those of traditional science labs. The artscience labs in Paris, Dublin, London, Boston, and Cambridge explore questions of relevance to contemporary society through experiments in art and design conducted within generally unfamiliar and often frontier areas of scientific knowledge. They can go beyond questions pursued by science labs to questions pertinent to society that nevertheless lack a clear scientific formulation. The issue of human identity, raised in a current exhibition at the Wellcome Collection, provides a good example. Who are we? The question, clearly pertinent, crumbles into an innumerable set of scientific questions, and many others a science lab can hardly pose today. But at Le Laboratoire, Shilpa Gupta and Mahzarin Banaji could explore the issue of terror and the subconscious mind, while Ryoji Ikeda and Benedict Gross could explore aesthetic interpretations of infinity. By avoiding definitive answers, artscience labs broaden the field of questions to be explored, and this exploration can lead to surprising discovery.

Through cultural exhibition, artscience labs provide a medium of public dialog that is both analogous to, and an alternative to, the medium of peer-reviewed publication in traditional science labs. The Science Gallery in Dublin and the Wellcome Collection in London exhibit experimental ideas touching on issues from infectiousness to obesity, and reach hundreds of thousands of gallery visitors each year. Visitors learn through the process, but so do the creative teams. Cultural exhibition, which is neither a proposition to be accepted or rejected nor a premise to be necessarily understood, opens up richer possibilities of expression. It also provides a more ready medium of public dialog than traditional peer-reviewed publications. The Dublin and London artscience labs record visitor number, age, profession, opinions, frequency of visitation, and other feedback that might possibly influence subsequent experiments. When Mathieu Lehanneur first exhibited the Bel-Air plant filter at Le Laboratoire, public reaction became the primary impetus for its later commercial development. The fact that data had not been collected to test the filter before exhibition, which would have dissuaded consideration in a strictly scientific (or commercial) context, actually enriched the meaning of the cultural experiment. The air filter was not presented as a work of science, to a science audience, but as a work of design to the general public. Mathieu would not have developed his plant filter in the way he did if he had been working in a conventional science lab. He developed it as cultural expression, without constraint, and this is how it was received. Only later did scientific testing show that the filter actually worked!

Artscience labs expose lab creation to a broader public than is possible with traditional science labs. Defined by their own creative process, science labs exist for a specialized audience of fellow scientists, while artscience labs coexist with the general public. The Science Gallery recently managed to make the science of bubbles—an arcane blend of mathematical and colloidal science within which I actually did my own doctoral research (very few noticed)—into a massively popular exhibition. Of course, this public orientation limits deep exploration, which takes place better in a science lab. Artscience lab research isn't a substitute for scientific investigation. But artscience labs make experimental investigation less isolating for creators and more comprehensible for the public, and the result is a kind of widening of the circle of the creative band. This allows more risky ideas to be carried further along and benefits innovation.

Artscience labs encourage a broad range of outcomes, including educational, cultural, humanitarian, and commercial. Creators pursue all kinds of dreams, not only those with commercial or cultural outcome. And often ideas, like Le Whif, move in wildly different directions, transforming their value proposition as they translate. This kind of broad interest is hard to find or encourage in traditional science labs, where funding is more tightly tied to scientific progress and industrial innovation. The idea translation lab at Harvard became popular when students realized that almost any idea, if hypothetically extrapolated to have large enough impact, would be encouraged within it. Creative students imagined dining halls that became highly interactive social networking nodes, catalyst companies for creative fashion design, music performances that connected to the rhythms of the beating hearts of audience members. Such rich environments favor innovation.

Artscience labs are small, young, and connected to institutions by the tethers of dreams. It is not necessary to have a doctoral degree to create or work on a creative team in an artscience lab. Moreover, the uncertainty and fast-changing nature of an artscience lab, the deep engagement in education, and the very nature of the cultural exhibition business all slant the organization toward attracting youth. Young, vibrant organizations exist all over the world, of course. This one funds experimentation in close association with large institutions that amplify the impact of the ideas translated. The Wellcome Collection is run by a team of thirteen, with an average age of thirty-four, and serves the role of innovation catalyst to the Wellcome Trust, Europe's largest foundation. The Science Gallery is run by a team of seven, with an average age of thirty-two, and serves the role of innovation catalyst to Trinity College, the oldest university in Ireland. The network of artsience labs I have described in this book, based in Cambridge, Boston, Paris, and Cape Town, employs a team of forty, with an average age of thirty-two, and aims to loosely serve the role of innovation catalyst to America's oldest university.

Artscience labs remain most useful when they retain an atmosphere of shared risk and ambition that guides creative bands in the pursuit of hypothetical ideas aimed at translational change. In this sense they are as fragile as are creators themselves.

Does this mean that prolific labs can exist for only brief periods, like the Bauhaus, or that they cannot exist independently of a major institution? Pablo Picasso and Thomas Edison remained productive throughout long careers, and the artscience lab network described in this book, while resonant with Harvard University, exists mostly independently of it. Anything is possible. But old age and institutional independence are strikes against artscience labs. We should not count on any one to last for very long.

≡8≡

TRANSLATIONAL CHANGE

Drawn by the mystery and intellectual dynamism at the frontiers of science, artists and designers are expanding the innovation boundaries of the traditional science lab. Through experimentation, they merge artistic and scientific processes in the same way that, in our own creative lives, we mix hypothesis with analysis, dreams with experimentation, when we translate any idea into an innovation.

The artscience lab curates this richly unpredictable process and guides it along, ultimately achieving sustainability through the occasional spinoff of cultural, humanitarian, and commercial products or startup companies. By facilitating innovations and then handing them off when the experimental phase is over, the artscience lab exchanges the social and cultural promise of early-stage innovations for the immediate resources needed to launch new experiments. This approach to sustainability, identical in spirit to that of a contemporary science lab, is the subject of this final chapter.

Achieving financial autonomy in a lab, even when you know the rules, is more laborious, tedious, and demanding of leadership than any other goal the lab takes on. Indeed, nothing may seem more challenging to leaders than this. They know that in a truly innovative lab, the drive for sustainability cannot lead them to sacrifice the spirit of experimentation. The successful lab can never be a purely commercial business, which sets some goal or receives some client directive—the creation of an object, or the solution to a certain problem—

and then, with fresh resources from clients or other investors, brings artists, scientists, and others together to translate one or more of these ideas into the requested outcome. This kind of business is obviously useful and may prove highly innovative, but it is not a lab. If the lab sacrifices the playful, contemplative, daring, irreverent atmosphere of the creative band in order to gain sustainability, the organization may survive, but the lab will cease to exist.

Consequently, the lab looks for ways to value the *process* of experimentation, not just the product. This experimental process, with its educational and cultural benefits, is inextricably linked to the "innovation bet" the lab makes on the creator when it undertakes any experiment, unable to foretell what valuable surprise, if any, will result. The lab's bet forms a bond of risk with the creator, who makes his or her own bet, and contributes to the energy, willingness, and shared vision of the creative band. What follows—experimentation, exhibition, commercialization—is a shared search not just for confirmation of the idea's value but also for mutual learning. The entire creative process, up to and through the demonstration of whatever social, humanitarian, or commercial value it may have, is the source of rich and transformative cultural dialog around a fascinating new idea whose financial risk, when it is acknowledged, serves to accentuate dialog and commitment.

Effective labs listen carefully to public reaction and translate what is learned into useful guidance for future experimentation. Creators, sensitive to the possibility that their work might not solicit interest among others in proportion to their own devotion, inevitably start to think differently about what they create when they begin to share it. And the best creators, from writers to theoretical particle physicists, listen, learn, and subsequently experiment differently with the new information that comes from wider exposure. Creator and lab move forward in this way, side by side, through a process of public dialog around the creative process.

In addition to advancing innovation, exhibition also produces extraordinary educational and cultural value. In its search for financial survivability, an effective lab manages to nurture its experimental spirit by acknowledging this added value, paying it serious attention, and procuring resources for its sustenance.

Each kind of artscience lab—educational, cultural, or translational—has, then, the opportunity to survive financially on the autonomous merits of its own experimentation. But each lab is also a part of the idea funnel, and benefits by the existence of its partner labs, which sometimes assume a larger share of the economic risk. Since each experiment in every lab cannot be expected to demonstrate to investors the full value that the creator has assigned by undertaking the experiment in the first place, each artscience lab need not demonstrate its own autonomous value all the time. This interdependence of the labs, mirroring the bond that holds together the creative band itself, helps assure sustainability and improves the willingness of artscience lab teams to take risks in what they do.

The Value of Learning: The Idea Translation Lab

The artscience lab is above all a learning environment. What, exactly, is learned in the lab, however, is not so easily measured, at least not in a standardized way. And as ideas translate, the value of this learning grows increasingly less obvious relative to the social and cultural value of the translated ideas themselves.

In the early days, as students imagined the object that later became the Pumpkin, they learned about cellular structure and the biological transport of water; and they learned about the challenges and health consequences of inadequate water transport in the developing world. Later, when the Pumpkin was fully designed and testing began in Africa, learning turned toward what effect the Pumpkin might have on ways of thinking about water transport, and on water transport itself. The Pumpkin emerged as an innovative idea in one kind of artscience lab environment, the educational artscience lab. Eventually, as the object came into existence, it evolved in cultural and commercial artscience labs. Learning happened at every point along the funnel, but it was especially obvious early on, when the cultural, humanitarian, and commercial values remained pure dreams.

The value of learning in the educational artscience labs is measured in terms of the following five key experiences:

◆ **Learning to matter:** To identify a hypothetical idea that is simultaneously a global need or opportunity and that matters personally to a passionate degree. This need might be purely cultural, humanitarian, or commercial, and generally addresses the human condition.

◆ **Learning to band:** To reach consensus within a group (usually three to five students with diverse skills and interests) around an original, hypothetical idea that can possibly address this need or opportunity, the translation of which will involve risks of sustained commitment that the group is willing to share.

♦ **Learning to learn:** To research the need, explore paths to idea translation, and conduct an experiment that tests a key hypothesis underlying the idea, an experiment that can be performed by the group within less than a year and with resources that the idea translation lab is able to provide.

♦ **Learning to persuade:** To present the idea translation as a proposal for funding in a way that is convincing first to peers, from whom one learns, then to a first public audience, from whom one learns even more.

♦ **Learning to act:** To work with the group and outside experts to perform the first idea translation experiment somewhere out in the world.

Of all the artscience labs, the educational artscience lab has the greatest creative impact. Not only do these idea translation labs bring fresh new minds into the idea funnel, they encourage the most surprising dreams, principally because the creators, being students, are unlikely to repeat what they have done before, and they have the least to lose through failure.

The first idea translation labs, created several years ago in Boston and Cambridge, today expose several thousand high school and university students each year to the principles of experiential learning through artscience creation. This happens through exhibition and other forms of idea participation, including events, workshops, and media coverage. To a more self-selecting group of several hundred students annually, the idea translation labs deliver the five core learning experiences described above. These labs also contribute innovative ideas that frequently continue to develop as they travel through the idea funnel. If they are eventually translated into viable commercial or humanitarian outcomes, they may not only produce a revenue stream but may help to drive corporate or institutional investment in artscience labs, in the same way that innovations emanating from conventional science labs drive investment in laboratory research.

In the first four years of the Harvard program, 120 students pursued twenty-four original ideas, including eleven projects undertaken by the 2009–2010 class. Eight of the thirteen pursued in the first three years are in serious development—more than half of the initial ideas contemplated in the idea translation class. In addition to MEND, Le Whif, MuseTrek, Lebone, the Pumpkin, and Andrea, these include Soccket, a soccer ball design that generates energy when it is kicked; Vertigrow, a design approach to vertical farming adapted to the needs of shantytown settings; and the Massachusetts biotech company Pulmatrix, which aims to commercialize a new broad approach to respiratory infectious disease. Around half of all students in the program have continued to pursue idea translation experiences after the class was over, and about a tenth of students have developed their ideas following graduation as founders and part-time or full-time employees of startup companies or artscience labs.

This track record helped the idea translation lab at Harvard receive consistent funding from university research and educational programs such as the Harvard Institute for Global Health, the Wyss Institute, and the university administration. It also became a close partner with the Graduate School of Design and the design program of the School of Engineering and Applied Sciences. Direct and indirect resources from these partners, together with endowment income, support annual running costs of the idea translation labs, including the direct and indirect costs of teaching, mentoring, project support, and travel.

Idea translation labs operate today in Boston, Oklahoma City, Dublin, Dhahran, Paris, and other cities. Many of these labs participate in the ArtScience Prize, which culminates in an innovation workshop at Le Laboratoire. There, idea translation lab students from around the world gather each summer and reinforce the five basic learning propositions of the idea translation labs in a rich multicultural context. This opportunity to realize dreams among a highly diverse student population provides a supplementary value to experiential learning. Cultural diversity shows up not just in Paris but in local programs that reflect the diversity of public education. In the 2009–2010 idea translation lab program in the Boston public schools, at the beginning of the year less than half of enrolled students (43 percent) reported speaking English at home. Other first languages included Spanish (18 percent), Vietnamese (8 percent), and Chinese (5 percent).

The Value of Exhibition: The Cultural Lab

As ideas move from the educational artscience lab to the cultural lab, they undergo another kind of experimentation: cultural exhibition. Like the value of educational experimentation, the value of cultural exhibition is most sensibly perceived over time. The difference is that with exhibition the experiment moves outside the mind of the creator and has the potential of spreading virally among the public. Even when the immediate impact of public exhibition is small (and this is often the case), what matters is that it starts a kind of kinetic movement of the idea in the minds of those who will carry it further, whether by invitation to exhibit in galleries and museums or by investment of outside resources to further experiment and develop the idea.

Value in cultural artscience labs is broadly measured in a couple of key ways:

◆ **Informed public interest:** First-time exhibition reaction is especially meaningful when the cultural artscience lab has developed a small, devoted, knowledgeable audience capable of providing the kind of learned feedback that peer review provides to science publication. This reaction is apparent in the published and online comments of journalists and other followers of cultural ideas, and is also perceived in the reaction of cultural professionals, from curators to potential commercial partners.

◆ **General public interest:** Experimental ideas reach the public by exhibition, usually starting with a small audience and progressively reaching wider audiences as the idea moves to larger venues. Numbers of visitors to exhibitions of original art and design ideas, and numbers of purchasers of original design products, are two meaningful measures of near-term impact, while long-term impact can be measured in terms of purchases of works of art or design for permanent collections and the integration of design ideas into sustainable commercial products.

In the first three years of our primary cultural artscience lab, Le Laboratoire, ten art and design experiments were performed and exhibited. Immediate press reaction averaged between twenty and thirty articles in the local and international print media, including art journals, business journals, educational publications, and mainstream media. To the degree that it was favorable, this reaction, in addition to Internet, radio, and television coverage, fanned partner interest and led to contracts for exhibition in larger galleries and museums around the world for every exhibition following the first. It also informed development of five original products: Le Whif, Andrea, MuseTrek, the Pumpkin, and (most recently) Edible Bottles.

Laboratoire exhibitions attracted approximately 15,000 visitors each year over the first three years of the organization, while works of art or design created at Le Laboratoire and exhibited in galleries or museums around the world annually attracted between 500,000 and a million visitors after the first year (these numbers include exhibitions of Laboratoire works in Brussels, New York, Hong Kong, Tokyo, Bangkok, Basel, Graz, and Copenhagen led by the artists and designers Mathieu Lehanneur, Shilpa Gupta, Ryoji Ikeda, James Nachtwey, and Marc Bretillot). Works of design (Andrea and Le Whif) reached approximately 50,000 people in 2009 through product sales and are estimated to reach between 500,000 and 1,000,000 people in Europe, Asia, and the Americas in 2010.

These outcomes have generated partner interest from a range of institutional collaborators, including France Télécom, Société Générale, BNP Paribas, Nestlé, Epson, LaboGroup, France Télévision, MoMA, the Centre Pompidou, Louisiana Museum of Modern Art north of Copenhagen, and the Louvre. These organizations have partnered with Le Laboratoire and helped generate resources to finance direct costs of experiments and exhibitions, less commonly as philanthropic supporters than as partners, renters of created works, or grant-makers aiming to catalyze innovation through experimentation. This partnership or grant-making parallels what takes place in the funding of science labs and reflects a belief in the value of experimentation on the part of the partner, rather than a contract for service or a demand for production.

These partnerships provide one-third of Le Laboratoire's funding. Another third of its total annual running costs are provided by the lower (commercial) end of the idea funnel (LaboGroup), and another third of the budget is provided by the higher end of the idea funnel (institutional supporters of the idea translation labs).

The Lab at Harvard provides another cultural laboratory model, where cultural value becomes apparent within an academic milieu. In its first year, The Lab at Harvard attracted around three thousand students, faculty, staff, and other visitors through exhibitions of original art and science ideas of Harvard students, works in progress that provided a forum of dialog and public exposure and promoted partnerships with diverse campus and community organizations. These included the American Repertory Theater, the Harvard Institute for

Global Health (HIGH), the Committee on African Studies, the Cambridge Science Festival, and Harvard's Visual and Environmental Sciences Department.

The Value of Innovative Products: The Translational Change Lab

The third kind of artscience lab, at the narrow end of the idea funnel, is the translational change lab. It aims to achieve sustainability of ideas that have emerged from artscience experimentation by developing them in commercial markets. These ideas remain experimental, and their commercial potential is still mostly secondary to their purely cultural value and social promise—otherwise they might better develop outside the lab. The translational change lab remains a lab even as it seeks to spin beneficial ideas out of the lab and into society.

Three obvious ways to measure the impact of the translational change lab are the following:

◆ **Prototype reaction and sales:** The translational change lab experiments first with public exhibition or sale of prototypes of experimental designs. Feedback on these designs is solicited in intimate environments intermediate between purely cultural and purely commercial or humanitarian-change environments, which permit the public to perceive, appreciate, and enter into the experimental creative process. Public feedback helps guide product development.

◆ **Commercial product reaction and sales:** After one or

more design iterations via lab experimentation, products enter into store and online sales channels. Product sales drive revenues, which help support the entire artscience lab network.

◆ **Culture and educational change:** As innovative products begin to sell, they change the culture and catalyze education within a broader public, with viral spread of new ideas (such as breathable food) mirroring the viral sale of the product.

The LaboGroup in Paris provides a working example of a translational change lab. It offers two environments for prototype testing: the FoodLab, where new food innovations are experienced through tastings and demonstrations, and the LaboShop, where prototype products are sold. Commercialization of its first products starts in select Paris stores, then moves to French and international markets. Launch of Andrea and Le Whif took roughly two years from conception to first commercialization. Having started to sell in the spring of 2009, Le Whif spun off as an independent London-based company, Breathable Foods, in late 2010. The Andrea technology was first sold from within LaboGroup in October 2009 and continues to develop there, with the possibility of becoming an independent company or being licensed to an existing commercial company.

Both products contribute to cultural change, in that we think differently about, and behave differently toward, the food we eat and the air we breathe. But both developed as fledgling ideas before the public eye, first through prototype sale in the translational lab's LaboShop, where the products received helpful feedback. Le Whif went through three different prototypes and a final commercial form while selling in the

LaboShop and by Internet. The translational lab sold hundreds of its first Le Whif prototype, thousands of its second, tens of thousands of its third, and in 2010 is projected to sell several hundred thousand of its commercial inhaler in stores internationally.

From the public perspective, the uniqueness of whiffing made experimenting even with a quirky early-stage product a cultural experience worth paying for, while for LaboGroup the ability to develop its product alongside the public helped the team overcome its limited commercial experience. This lack of knowledge was actually probably an advantage—the product itself being so new that significant commercial experience with the food industry might have dissuaded the team from betting on something so unusual.

Leadership and Funding

An efficient association of interests among the various art-science labs can help sustain ideas as they move through the idea funnel. This happens naturally through commercial markets in other idea funnels, such as Silicon Valley. An idea published out of a university lab may grab the attention of a local venture capitalist, who sees market potential and invests to create a startup. If the idea proves successful in the applied research environment of the startup and generates sufficient media interest, it may move on through sale of the startup to a fully commercial company in the Valley or somewhere else. Commercial markets provide the gravity that accelerates ideas as they move down the funnel.

In artscience labs, where the aims of idea development are cultural or humanitarian betterment rather than purely commercial profit, another kind of guiding force is needed. To preserve the simultaneous educational, cultural, humanitarian, and commercial interests of the various labs in the funnel, there must be some means of decision-making and dialog where all interests are represented. This association of interests takes the form of a single hand guiding—sometimes very lightly, at other times more forcefully—ideas from top to bottom of the funnel. To achieve such cooperation, the artscience lab network described in this book is managed by ArtScience Labs, an organization that gives directors from each lab a voice in the network's operation.

Each spring, the educational, cultural, and translational lab directors meet and examine ideas that have emerged from educational labs and will be brainstormed and further developed in the summer workshop in Paris. The group discusses the ideas, creators, and teams behind the ideas, and possible paths these ideas might follow through the idea funnel. Projects with cultural exhibition potential may appeal to Le Laboratoire in Paris; others with commercial potential may appeal to LaboGroup.

Each summer, the leadership of ArtScience Labs gets together to discuss ideas leaving the funnel as sustainable commercial, humanitarian, and cultural engagements, resources needed for the coming year in any of the labs, personnel issues, partnering possibilities, operational challenges, and long-term strategy.

And each fall the same directors discuss ideas moving from Le Laboratoire into LaboGroup or somewhere else. These conversations relate not just to the commercial and humanitarian impact of ideas but the educational and cultural missions to which the ideas have, until then, been aimed. As the funnel narrows from top to bottom, it becomes critical for the future of the labs that participants at the wider end of the funnel who are not engaged in translational change at the narrow end—from students to members of creative bands—be kept informed about the progress of their lab's ideas as those ideas move into the larger world.

Each year, ArtScience Labs aims to involve up to three hundred students, produce up to three exhibitions as outcomes of major artist and designer-led experiments, and launch at least one new product. ArtScience Labs also helps facilitate the creation and startup of other labs outside the network, less through consulting services than through common experimental projects that help build experience and infrastructure. These kinds of partnerships exist in Singapore, with the St. Joseph's Institution, which is seeking to start an Art-Science Prize; in Oklahoma City, which is also seeking to build an ArtScience Prize and idea translation lab program; and in Dhahran, Saudi Arabia, with the support of Aramco, which is seeking to animate the core innovation program of the King Abdulaziz Center for World Culture.

Managing the idea funnel in this way opens the possibility of attracting "cultural venture capital," meaning investment that aims to produce solutions to problems of society through cultural engagement. Money invested in the idea funnel, assuming it possesses a transparent management structure, can be distributed among the various labs of the funnel, producing educational, cultural, and financial returns on investment, with an overall objective of returning the principal within a several-year horizon. Cultural venture capital largely drove the startup of the artscience lab network described in this book.

Sometimes money returned to cultural venture capitalists is gifted back to the artscience labs' endowment program, a second source of funding dedicated to the educational mission of the labs. Endowment resources drive all educational activities in the Boston and Cambridge artscience labs, and many educational activities in other labs of the funnel. At heart, the artscience lab network deliberately blurs the distinction between creation and education, just as it blurs the distinction between art and science, artist and designer, engineer and artisan. Artscience labs may sometimes contribute to beneficial societal change, but they should always serve to motivate young creators to dream and to learn through pursuit of dreams. Endowment resources help ensure that education continues even while innovation and translational change remains a promise.

A final source of funding for artscience labs comes through revenues generated by commercial activities. Product sales, which may also pay royalties to students, designers, and artists involved in the creative process, contribute to financing new innovations in cultural, industrial, and humanitarian contexts and can eventually contribute to the educational endowment.

With effective oversight, the network of artscience labs described in this book has great financial and administrative flexibility to invite open-ended experimentation, to support creators in the iterative process of innovation, and to maintain a rich dialog with a public that welcomes surprise.

The Bauhaus

Walter Gropius formed the Bauhaus in the spring of 1919 in the German town of Weimar. Germany had just lost the war, the economy was in ruins, citizens had been bombed, shot, gassed, and maimed. What had it all been worth? It seemed clear to war-scarred artists such as Gropius that art had served the wrong cause.

As described in his manifesto and the *Program on the State of the Bauhaus in Weimar*, the Bauhaus aimed to rescue art from its prewar isolation. It would elevate nonappreciated artistic activities, like woodcutting, to the status of the fine arts, build links with industry, and seek financial autonomy through the sale of products. Creation, not theory, would be the basis of education, and the artist would be an experimentalist.

Gropius's school grew—influenced by the teachings of nineteenth-century anti-industrialists John Ruskin and William Morris, built on an early twentieth-century industrialist-funded organization called the Werkbund, and later inspired by designer Peter Behrens and painter, designer, and architect Henry van de Velde. Seizing its small window of opportunity, the Bauhaus changed the course of Western design and architecture, as magnificently illustrated in a recent MoMA catalog *Bauhaus 1919–1933: Workshops in Modernity*. Its experience, designs, and architecture inspired the young New York Museum of Modern Art, was a model for art and design education programs around the world, and in many ways changed how homes, kitchens, offices, buildings, and furniture would be made for a century.

Today, arguably, the coincidences of technological revolution and global economic crises make our time as challenging for large institutions as the postwar years were for the Weimar Republic. It is easy to agree that the old institutional paradigms—those that guided us from the world wars into the twenty-first century—are ill-suited to the conditions faced by the present post-Google generation: a planet that is heating up, social safety nets stretched to the breaking point or already in tatters, and a workforce that is too specialized to dream credibly and sustainably of a more promising future. It is less clear what kind of opportunities we have to address these critical needs.

Hoping to change things for the better, we have long turned to labs. The model of the science lab—how it recruits experimentalists, what these experimentalists do, how they communicate and eventually translate what they do, and how they receive financial support for their experimentation—is by now relatively well understood. The artscience labs described in this book provide a new model, though in some ways it is as old and intuitive as the model by which we build environments and networks of friends who support our dreams. This role of intuition is essential to the added value of the artscience lab.

Of course, tethering our educational, cultural, humanitarian, and commercial future to something as unpredictable as the creative process itself may appear foolhardy. Most of us, now so far removed from the innovative spirit we knew as children or young adults, forget how invigorating it is to pursue hypothetical ideas of our own making that remain as yet unrealized. Cherishing what happens at the end of a productive adventure, we see too clearly, in hindsight, the very real risks we ran that it might not have happened at all. Sometimes with a shudder at how close we came to failure, we have difficulty, in this post-dream state, conjuring up the *excitement* that comes from knowing our ideas may not be realized, depending on what we do. This spirit of real risk-taking in pursuit of a shared passion is what animates the artscience lab. It makes us wish to participate, to collaborate without coercion. Our days come to matter all the more, because we recognize that our dreams might not matter without us.

Acknowledgments

A book like this, written about an experience one is living, an experiment within an experiment, would be incomprehensible without the intelligence and commitment of fabulous editors, and I have had two at Harvard University Press: Michael Fisher, who led the editing of this book, hammer on a rough rock; and Susan Wallace Boehmer, who applied the polish. I wish also to thank many readers of various parts of the English and French versions, including Michael John Gorman, Jay Cantor, Peter Galison, Ken Arnold, Valérie Abrial, Olivier Borgeaud, Xavière Masson, Caroline Naphegyi, my agent Ike Williams, and my wife, Aurélie.

Index

Adrià, Ferran, 34, 116
AIDS, 33, 43, 137
Air filter. *See* Bel-Air/Andrea
Albers, Josef, 154
Allen, Susie, 53
Altruism/humanitarianism, 59, 134, 135, 136, 139, 149, 184. *See also* Lebone; Medicine in Need (MEND); Pumpkin, the
Amar, Laurent, 112, 113
Andrea. *See* Bel-Air/Andrea
Arbus, Cashman, 129
Arnold, Ken, 8, 9–10, 79
Ars Electronica Futurelab, 8, 21, 23, 153
Art, 17, 21, 23, 80–81, 83, 103
Art & Sciences Collaborations, Inc. (ASCI), 9
Artbots, 103
Artscience, 16, 160
Artscience Lab (organization), 192–194
Artscience lab(s), 17, 35, 181, 194; art and science in, 8; and Bauhaus, 8, 196; and collaborative humanitarian models, 136; commercial, 11; and conventional science labs, 172, 173–175; and creative bands, 153, 159–160, 172; and creative process, 45–46; and economic risk, 181; funding for, 191–194; and idea funnel, 11; and innovation, 8, 130; investment

value of, 14; and multilab structure, 8; network of, 23, 26, 37, 63, 190, 192; and public dialog, 11; and public exposure, 174; range of outcomes from, 174–175; and risk, 26; and social and cultural change, 11; support of creators by, 13–14; and sustainability, 179; and youth, 175–176
ArtScience Prize, 67, 69, 71, 73, 128, 185, 193
Arts Task Force, 64
Azambourg, François, 43, 148

Banaji, Mahzarin, 37, 96–100, 101, 172
Bargh, John, 98
Bauhaus, 8, 154, 176, 195–197
Behrens, Peter, 195
Bel-Air/Andrea, 33, 110–116, 184; commercial development of, 35, 36, 110–116, 117, 118, 173–174, 190; commercial need for, 126; and public exhibition, 187; public interest in, 33; revenue from, 42
Benayoun, Julien, 144–145, 147
Bencherif, Sidi, 148
Bibette, Jérôme, 34–35, 116, 117, 164
Bill & Melinda Gates Foundation, 17, 33–34, 42, 43, 139, 141
Bloom, Barry, 146
Blumenthal, Heston, 34
Borgeaud, Olivier, 31, 109
Boston Center for the Arts, 66

Index

Index